Octanol–Water Partition Coefficients: Fundamentals and Physical Chemistry

Wiley Series in Solution Chemistry

Editor-in-Chief

P. G. T. Fogg, *University of North London, UK*

Editorial Board

Volume 1
pH and Buffer Theory — A New Approach
H Rilbe
Chalmers University of Technology, Gothenburg, Sweden

Volume 2
Octanol–Water Partition Coefficients: Fundamentals and Physical Chemistry
J Sangster
Sangster Research Laboratories, Montreal, Canada

Octanol–Water Partition Coefficients: Fundamentals and Physical Chemistry

J. Sangster

Sangster Research Laboratories, Montreal, Canada

Wiley Series in Solution Chemistry
Volume 2

JOHN WILEY & SONS

Chichester·New York·Weinheim·Brisbane·Singapore·Toronto

Other Wiley Editorial Offices

John Wiley & Sons, Inc., 605 Third Avenue,
New York, NY 10158–0012, USA

VCH Verlagsgesellschaft mbH,
Pappelallee 3, D–69469 Weinheim, Germany

Jacaranda Wiley Ltd, 33 Park Road, Milton,
Queensland 4064, Australia

John Wiley & Sons (Asia) Pte Ltd, 2 Clementi Loop #02–01,
Jin Xing Distripark, Singapore 129809

John Wiley & Sons (Canada) Ltd, 22 Worcester Road,
Rexdale, Ontario M9W 1L1, Canada

British Library Cataloguing in Publication Data

A catalogue record for this book is available from the British Library

ISBN 0–471–97397 1

Typeset in 10/12pt Times by Dobbie Typesetting Limited, Tavistock, Devon
Printed and bound in Great Britain by Biddles Ltd., Guildford, Surrey
This book is printed on acid-free paper responsibly manufactured from sustainable forestation, for which at least two trees are planted for each one used for paper production.

Contents

Series Preface

There are many aspects of solution chemistry. This is apparent from the wide range of topics which have been discussed during the International IUPAC Conferences on Solubility Phenomena. The Wiley Series in Solution Chemistry was launched to fill the need to present authoritative, comprehensive and up-to-date accounts of these many aspects. Internationally recognized experts from research or teaching institutions in various countries have been invited to contribute to the Series.

Volumes in print or in preparation cover experimental investigation, theoretical interpretation and prediction of physical chemical properties and behaviour of solutions. They also contain accounts of industrial applications and environmental consequences of properties of solutions.

Subject areas for the Series include: solutions of electrolytes, liquid mixtures, chemical equilibria in solution, acid–base equilibria, vapour–liquid equilibria, liquid–liquid equilibria, solid–liquid equilibria, equilibria in analytical chemistry, dissolution of gases in liquids, dissolution and precitipitation, colubility in cryogenic solvents, molten salt systems, solubility measurement techniques, solid solutions, reactions within the solid phase, ion transport reactions away from the interface (i.e. in homogeneous, bulk systems), liquid crystalline systems, solutions of macrocyclic compounds (including macrocyclic electrolytes), polymer systems, molecular dynamic simulations, structural chemistry of liquids and solutions, predictive techniques for properties of solutions, complex and multi-component solutions, applications of solution chemistry to materials and metallurgy (oxide solutions, alloys, mattes..), medical aspects of solubility, and environmental issues involving solution phenomena and homogeneous component phenomena.

Current and future volumes in the Series include both single authored and multi-authored research monographs and reference level works as well as edited collections of themed reviews and articles. They all contain comprehensive bibliographies.

Volumes in the Series are important reading for chemists, physicists, chemical engineers and technologists as well as environmental scientists in academic and industrial institutions.

May 1996   Peter Fogg

CHAPTER 1

Introduction

1 THE PARTITION COEFFICIENT

Many organic liquids are more or less immiscible with each other or with water at ordinary temperatures and pressures (Francis, 1961). If a third substance be added to a system of two immiscible liquids, the added component will tend to distribute itself between the two solvents until, at equilibrium, the ratio of the concentrations (or mole fractions) of the distributed substance will attain a certain value (Figure 1.1). This ratio, and its relative insensitivity to variations in temperature and concentration, was noted and first studied quite early (Berthelot and Jungfleisch, 1872; Nernst, 1891). For a solute X distributed between two immiscible solvents I and II, the ratio

$$[X]_{II}/[X]_{I} = \text{constant} \tag{1.1}$$

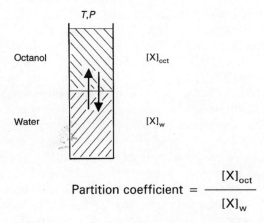

Figure 1.1 Definition of the partition coefficient

came to be called the partition or distribution ratio (or constant or coefficient). Its relative insensitivity to concentration and temperature led to the use of the terms 'distribution law' or 'partition law'. The term in French was 'coefficient de partage' (Berthelot and Jungfleisch, 1872) but is now more commonly 'coefficient de répartition'. The German 'Teilungskoeffizient' (Nernst, 1891) is now more commonly 'Verteilungskoeffizient'.

2 EFFECT OF CONCENTRATION

Careful study (Berthelot and Jungfleisch, 1872; Nernst, 1891) revealed that, in some cases, the distribution ratio, Eq. (1.1), was clearly concentration dependent. In the distribution of benzoic acid between benzene and water, for example, it was found that a modified ratio

$$\sqrt{[X]_{II}}/[X]_I = \text{constant} \tag{1.2}$$

was the proper expression of the distribution law (benzene was solvent II, water I). In this case, the dimerization process

$$2C_6H_5COOH \rightarrow (C_6H_5COOH)_2 \tag{1.3}$$

is taking place in the benzene layer. Alternatively, if a substance is ionized in aqueous solution, then this fact also can be taken into account in formulating the distribution law (Leo et al., 1971; Craig and Craig, 1950). Though the behaviour represented by Eq. (1.2) might illuminate the process of association taking place in phase II, from the point of view of consistency of interpretation it is preferable to define the partition coefficient as referring to the same molecular species in both solvents, as Nernst did (Nernst, 1891). This provision is now part of the definition of the true partition coefficient.

Even if the distributed substance does not associate or dissociate in the solvents, there will in general be a concentration dependence of the partition coefficient as defined by Eq. (1.1). This is due to the fact that, as concentration is increased, the solute passes from the dilute solution region to the concentrated solution region., i.e., beyond the region where Henry's law holds. Outside the dilute solution region, solute–solute interactions become progressively more important and cannot be neglected. This is represented schematically in Figure 1.2, which shows a liquid–liquid equilibrium triangular diagram of the usual type. Here the solute is completely miscible with both solvents which themselves are immiscible. (Corresponding diagrams could be drawn for the cases in which the solute is slightly soluble in either or both solvents. The same conclusions would hold in these cases, but would not be as clearly seen diagrammatically.) Eq. (1.1) corresponds to the case in which the ratio of concentrations of solute (x, x') at the ends of the

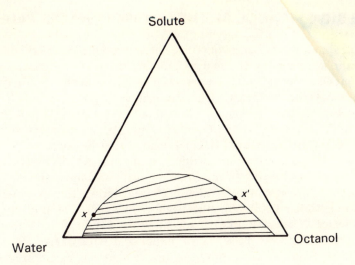

Figure 1.2 Distribution of a solute between octanol and water. The miscibility between the solvents is exaggerated for clarity

tie lines is constant in the two-phase region. Clearly, in the general case this is not true. Eq. (1.1) holds, however, if $[C] \to 0$, i.e., where the solutions are sufficiently dilute. This condition is now part of the definition of the true partition coefficient.

3 HYDROPHOBIC OR HYDROPHILIC?

Most organic compounds are miscible with water to only limited extent (Francis, 1961; Lide, 1992; Sørenson and Arlt, 1979) at ordinary temperatures and pressures. This fact may be accounted for on several levels of explanation ('like dissolves like', solution thermodynamics, intermolecular forces...). Organic compounds are correspondingly commonly soluble in solvents such as chloroform, diethyl ether and benzene. Water, a polar solvent, will then tend to be a good solvent for polar solutes, i.e., solutes which contain functional groups such as $-OH$, $-CHO$, $-COOH$, $-NO_2$, $-NH_2$. Molecules containing only carbon and hydrogen are said to be non-polar. Organic compounds can thus be described as more or less hydrophobic (lipophilic) or hydrophilic (lipophobic). A *lipid* is a fat or fat-like substance. As described below, for certain purposes a quantitative measure of a substance's hydrophilic or lipophilic nature is very useful. Water solubility, a commonly-measured property of organic compounds, is sometimes used as such a quantitative measure. For historical and other reasons, the octanol–water partition coefficient is even more extensively used for such purposes.

4 THE SIGNIFICANCE OF THE OCTANOL–WATER PAIR

As stated above, it was Berthelot (Berthelot and Jungfleish, 1872) and Nernst (1891) who were among the first to examine the partition coefficient as a purely physicochemical phenomenon. The early use of this property in the correlation of the biological action of simple organic solutes has been sketched by Rekker (Rekker and Mannhold, 1992) and Hansch and Leo (1995). Both Meyer (1899) and Overton (1901, 1991) demonstrated that the narcotic action of many organic compounds could be more or less closely correlated with their oil–water partition coefficients. At this early date, 'oil' meant very often olive oil, or any common water-insoluble organic solvent. Collander made systematic partition coefficient measurements using isobutanol–water (Collander, 1950), ether–water (Collander, 1949), oleyl or octyl alcohol–water (Collander, 1951) solvent pairs, and related ether–water partition coefficient by the work of Fujita and co-workers (Hansch and Fujita, 1964; Fujita *et al.*, 1964) who showed that this partition coefficient could provide a rationalization for the interaction of organic compounds with living organisms or for biological processes occurring in organisms. (The electronic Hammett parameter σ and Taft steric parameter E_s were combined with the octanol–water partition coefficient in the form of a 'hydrophobic parameter' π, but these details are less important for the present exposition.)

5 SYMBOLS AND NOMENCLATURE

Several symbols are in common use for the octanol–water partition coefficient: D, K, P and V (Verteilungskoeffizient). For the true partition coefficient (same molecular species in both solvents, dilute solutions), 'P' and 'K_{ow}' are used. K_{ow} will be used henceforth in this book. As a general rule, P is preferred by medicinal and pharmaceutical chemists and K_{ow} is used most often by environmental and toxicological chemists. 'D' (distribution coefficient) is usually reserved for partition coefficients measured under conditions in which the solute is partially or completely ionized in the aqueous phase or which otherwise do not correspond to the definition of true partition coefficient. The distribution coefficient, as here described, is sometimes called the apparent partition coefficient (P_{app}).

6 PROPERTIES OF 1-OCTANOL AND WATER

1-Octanol is a long-chain normal alcohol ($CH_3CH_2CH_2CH_2CH_2CH_2CH_2CH_2OH$) and thus contains both a hydrophobic hydrocarbon chain and a hydrophilic end group. It may be imagined to approximate, more faithfully than a purely non-polar solvent (CCl_4, hexane, etc.) the physicochemical environment experienced by a test chemical in living tissue. Some properties of octanol and water are presented in Table 1.1. Cyclohexane, a common hydrocarbon solvent, is added for comparison.

Like many organic solvents, octanol is less dense than water, and, while normally considered immiscible with water (the critical solution temperature for octanol and water (Francis, 1961) lies above 265°C), this is not strictly true. Data on the mutual solubility of octanol and water are presented in Tables 1.2 and 1.3. Selected smoothed data are shown in Figures 1.4 and 1.5. The solubility of water in octanol is surprisingly high (2.2±0.2 M at 25°C).

At 25°C the solubility of water in octanol may be taken to be 0.275 mole fraction (Apelblat, 1983) and for octanol in water, 7.5×10^{-5} mole fraction (Marcus, 1990). The corresponding data for octane and water are (Sørensen and Arlt, 1979) 8.1×10^{-4} and 1.1×10^{-7}. Evidently the presence of the hydroxyl group in the octanol molecule greatly enhances the mutual solubilities. According to Dearden and Bresnen (1988), the solubility of water in octanol passes through a maximum, and the solubility of octanol in water passes through a minimum. Other results (von Erichsen, 1952; Stephenson *et al.*, 1984; Dallos and Liszi, 1995) suggest that the solubilities of both solvents increase monotonically in the experimental temperature range (87.9 at 0°C, 55.5 at 100°C), water in effect becomes less polar as the temperature increases, and might be expected to be a better solvent for octanol at higher temperatures.

7 THE STRUCTURE OF 'WET' OCTANOL AND IMPLICATIONS

A saturation mole fraction of water in octanol of 0.275 means that, on average, for every 100 molecules of octanol there are 38 water molecules present. This

Table 1.1 Selected properties of 1-octanol, water and cyclohexane (Lide, 1992; Liew *et al.*, 1992; Berti *et al.*, 1986)

Property	Temperature (°C)	Unit	1-octanol	Water	Cyclohexane
Molecular weight		$g\,mol^{-1}$	132.2	18.02	84.2
Density	25	$g\,cm^{-3}$	0.8223	0.9969	0.7785
Molar volume		$cm^3\,mol^{-1}$	160.8	17.96	108.2
Melting point		°C	−15.4	0.0	6.5
Normal boiling point		°C	195.2	100.0	80.7
Vapour pressure	25	kPa	0.0145	3.17	13.0
Heat capacity	25	$J\,mol^{-1}\,K^{-1}$	317.5	75.3	154.9
Dielectric constant	20		10.3	80.2	2.02
Refractive index	20		1.4295	1.333	1.4266

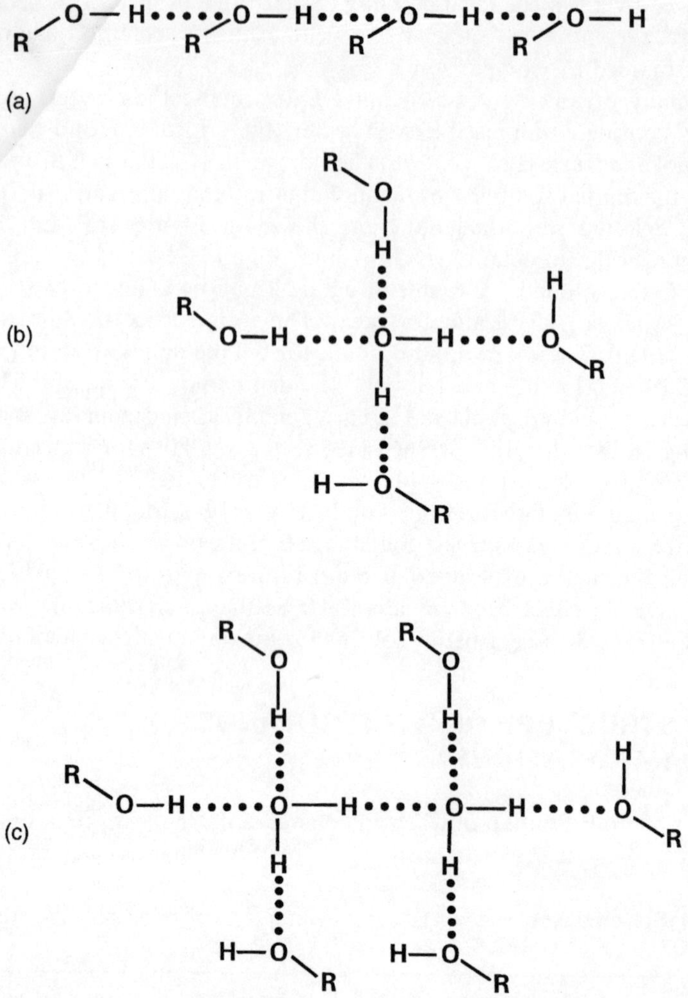

(a)

(b)

(c)

Figure 1.3 Schematic representation of liquid structure: solid lines covalent bonds; dotted lines hydrogen bonds. (a) Pure 1-octanol. (b) Water at infinite delution in 1-octanol. (c) 75% saturation of water in 1-octanol (Marcus, 1990)

rather large solubility might give one pause, but in fact the solubility of water in C_4–C_8 alkanols is in the range 0.26–0.68 mole fraction at 25°C (Apelblat, 1983). Thermodynamic (Marcus, 1990) and theoretical (DeBolt and Kollman, 1995) considerations indicate that 'wet' octanol is actually microheterogeneous in structure. Marcus (1990) combined experimentally-determined excess Gibbs energies of water-in-octanol solutions (Apelblat, 1983) with a 'quasi-lattice quasi-chemical' model of wet octanol. He deduced that, although the bulk mole

Table 1.2 Solubility of 1-octanol in water*

Temperature (°C)	Octanol mole fraction ($\times 10^5$)	Reference
5	5.60	Lebedinskaya et al. (1984)
5	16.1	Dearden and Bresnen (1988)
10	5.50	Lebedinskaya et al. (1984)
15	5.40	Dallos and Liszi (1995)
15	7.42	Vochten and Petre (1973)
20	5.73	Addison (1945)
20	5.40	Lebedinskaya et al. (1984)
20	5.60	Dallos and Liszi (1995)
20	6.25	Sørenson and Arlt (1979)
20.5	6.68	Stephenson et al. (1984)
25	8.00	Butler et al. (1933)
25	6.84	Crittenden and Hixon (1954)
25	5.90	Dallos and Liszi (1995)
25	11.9	Dearden and Bresnen (1988)
25	7.36 (R)	Hefter (1984)
25	6.68	Kinoshita et al. (1958)
25	8.05	McBain and Richards (1946)
25	6.88	Shinoda et al. (1959)
25	7.03 (R)	Sørenson and Arlt (1979)
25	9.32	Dallas and Carr (1992)
30	6.20	Dallos and Liszi (1995)
30	11.4	Dearden and Bresnen (1988)
30	13.7	Rao et al. (1961)
30	10.5	Sobotka and Glick (1934)
30.6	8.73	Stephenson et al. (1984)
35	6.50	Dallos and Liszki (1995)
35	10.9	Dearden and Bresnen (1988)
40	6.90	Dallos and Liszi (1995)
40	10.2	Dearden and Bresnen (1988)
40	7.20	Lebedinskaya et al. (1984)
40	8.18	Lavrova and Lesteva (1976)
40	14.4 (R)	Sørenson and Arlt (1979)
40.1	8.87	Stephenson et al. (1984)
45	7.40	Dallos and Liszi (1995)
45	9.83	Dearden and Bresnen (1988)
50	7.80	Dallos and Liszi (1995)
50	9.12	Dearden and Bresnen (1988)
50	14.3	Stephenson et al. (1984)
55	9.51	Dearden and Bresnen (1988)
60	8.18	Lavrova and Lesteva (1976)
60	21.1 (R)	Sørenson and Arlt (1979)
60	9.00	Lebedinskaya et al. (1984)
60.3	12.0	Stephenson et al. (1984)
70	15.9	Dearden and Bresnen (1988)
70.3	10.5	Stephenson et al. (1984)
80	12.6	Lebedinskaya et al. (1984)
80.1	11.9	Stephenson et al. (1984)
90.3	11.7	Stephenson et al. (1984)
95	137.	Zhuravleva et al. (1977)

*(R) indicates a value recommended by author cited. Three significant figures are given regardless of citation.

Table 1.3 Solubility of water in 1-octanol[a]

Temperature (°C)	Mole fraction of water	Reference
0	0.203	von Erichsen (1952)
5	0.288	Lebedinskaya *et al.* (1984)
5	0.206	Dearden and Bresnen (1988)
7	0.158	Zhuravleva *et al.* (1977)
10	0.224	von Erichsen (1952)
10	0.251	James (1983)
10	0.284	Lebedinskaya *et al.* (1984)
10	0.241	Stephenson *et al.* (1984)
15	0.273	Beezer *et al.* (1980)
15	0.208	Brodin *et al.* (1976)
15	0.271	Dallos and Liszi (1995)
15	0.265	Tokunaga *et al.* (1980)
19	0.250	Stephenson *et al.* (1984)
20	0.273	Dallos and Liszi (1995)
20	0.245	von Erichsen (1952)
20	0.257 (R)	Hefter (1984)
20	0.284	Lebedinskaya *et al.* (1984)
20	0.260	James (1983)
20	0.194 (R)	Sørenson and Arlt (1979)
20	0.268	Tokunaga *et al.* (1980)
22	0.264	Kristl and Vesnaver (1995)
23	0.210	Zhuravleva *et al.* (1977)
25	0.263	Ababi and Popa (1960)
25	0.278	Apelblat (1983)
25	0.272	Beezer *et al.* (1980)
25	0.223	Brodin *et al.* (1976)
25	0.283	Dearden and Bresnen (1988)
25	0.289	Dallas and Carr (1992)
25	0.261 (R)	Hefter (1984)
25	0.113	Crittenden and Hixon (1954)
25	0.274	Dallos and Liszi (1995)
25	0.207 (R)	Sørenson and Arlt (1979)
25	0.269	Tokunaga *et al.* (1980)
30	0.276	Dallos and Liszi (1995)
30	0.291	Dearden and Bresnen (1988)
30	0.265	von Erichsen (1952)
30	0.266 (R)	Hefter (1984)
30	0.281	James (1983)
30	0.261	Rao *et al.* (1961)
30	0.270	Tokunaga *et al.* (1980)
30.5	0.256	Stephenson *et al.* (1984)
35	0.238	Brodin *et al.* (1976)
35	0.278	Dallos and Liszi (1995)
35	0.296	Dearden and Bresnen (1988)
35	0.273	Tokunaga *et al.* (1980)
35	0.234	Zhuravleva *et al.* (1977)
40	0.272	Beezer *et al.* (1980)

Table 1.3 *(continued)*

Temperature (°C)	Mole fraction of water	Reference
40	0.240	Brodin *et al.* (1976)
40	0.281	Dallos and Liszi (1995)
40	0.296	Dearden and Bresnen (1988)
40	0.290	von Erichsen (1952)
40	0.289	Lebedinskaya *et al.* (1984)
40	0.283 (R)	Hefter (1984)
40	0.298	James (1983)
40	0.283	Lavrova and Lesteva (1976)
40	0.247 (R)	Sørenson and Arlt (1979)
40	0.270	Stephenson *et al.* (1984)
40	0.276	Tokunaga *et al.* (1980)
45	0.283	Dallos and Liszi (1995)
45	0.297	Dearden and Bresnen (1988)
50	0.286	Dallos and Luszi (1995)
50	0.267	Dearden and Bresnen (1988)
50	0.312	von Erichsen (1952)
50	0.275	Stephenson *et al.* (1984)
52	0.279	Zhuravleva *et al.* (1977)
55	0.257	Dearden and Bresnen (1988)
60	0.333	von Erichsen (1952)
60	0.327 (R)	Hefter (1984)
60	0.293	Lebedinskaya *et al.* (1984)
60	0.319	Lavrova and Lesteva (1976)
60	0.292 (R)	Sørenson and Alrt (1979)
60.2	0.292	Stephenson *et al.* (1984)
70	0.356	von Erichsen (1952)
70	0.202	Dearden and Bresnen (1988)
70.1	0.296	Stephenson *et al.* (1984)
74	0.319	Zhuravleva *et al.* (1977)
80	0.376	von Erichsen (1952)
80	0.297	Lebedinskaya *et al.* (1984)
80.1	0.294	Stephenson *et al.* (1984)
90	0.397	von Erichsen (1952)
90.5	0.301	Stephenson *et al.* (1984)
100	0.420	von Erichsen (1952)

[a](R) indicates a value recommended by author cited. Three significant figures are given regardless of citation.

fraction of water at saturation at 25°C is 0.275, the *local* mole fraction of water around a water molecule is considerably greater (0.472) and that around an octanol molecule, somewhat less (0.200).

DeBolt and Kollman (1995) performed computer molecular dynamics calculations for pure 1-octanol and its water-saturated solution using an optimized intermolecular potential function for liquid simulations. Various structural, dynamic and energetic properties of these systems were calculated

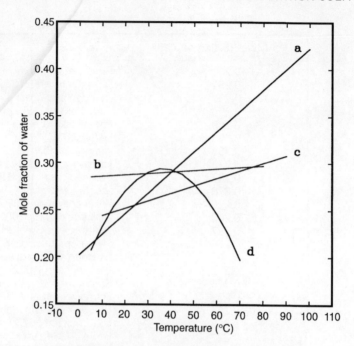

Figure 1.4 The solubility of water in 1-octanol as a function of temperature (smoothed data): a, von Erichsen (1952); b, Lebedinskaya *et al.* (1984); c, Stephenson *et al* (1984); d, Dearden and Bresnen (1988)

and compared with experiment. For present purposes, it is sufficient to note here that they confirmed the general conclusions of Marcus (1990). These may more easily be grasped by considering a diagrammatic representation of liquid structure (Figure 1.3). In pure (absolutely dry) 1-octanol, there is cooperative hydrogen bonding among octanol molecules (Figure 1.3a). Both rings and extended chains can exist; DeBolt and Kollman (1995) calculated the frequency distribution of octanol cluster sizes at 40° and 75°C. At infinite dilution in octanol (Figure 1.3b), the water molecule is surrounded tetrahedrally by four octanol molecules. At 75% saturation (Figure 1.3c), the water molecules are hydrogen bonded both to each other and to octanol molecules. In all cases, it is seen that the hydrocarbon tails of solvent molecules often are contiguous; the result is a tendency to form small inverted micellar-like regions (DeBolt and Kollman, 1995). Of course, in real time such hydrogen-bonded aggregates are extremely short-lived, but re-form in equivalent fashion, similar to the 'flickering' cluster model of water structure (Frank, 1972).

This picture of microheterogeneity in wet octanol has received further support from experimental data on the dependence of heat of solution of organic compounds in octanol as a function of water content (Bernazzani *et*

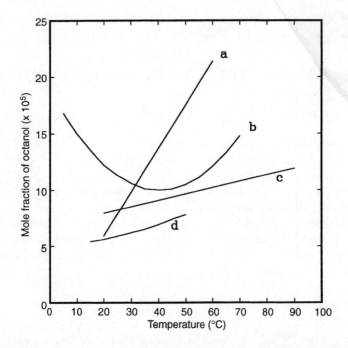

Figure 1.5 The solubility of 1-octanol in water as a function of temperature (smoothed data): a, Sørenson and Arlt (1979) 'recommended'; b, Dearden and Bresnen (1988); c, Dallos and Liaszi (1995); d, Stephenson et al. (1984)

al., 1995). Experimental and theoretical considerations suggest that, at about $x_W = 0.07$, water molecules in the inverted micellar-like regions begin to form water–water hydrogen bonds. This behaviour increases with water content up to the saturation limit.

At first sight, octanol may seem to be an isotropic bulk solvent of low dielectric constant, essentially hydrophobic or 'oily'. Detailed considerations (Marcus, 1990; DeBolt and Kollman, 1995) have shown that octanol itself may be microheterogeneous through hydrogen bonding. In the classical shake-flask definition of the partition coefficient (Figure 1.1), moreover, octanol becomes water-saturated to the extent of 27.5 mole %. It thus becomes a partitioning solvent quite different from hydrocarbons or other aprotic solvents. It has in fact been found that correlations between biological activity and partition coefficients are distinctly more successful for octanol–water than for, say, hydrocarbon–water pairs (Smith *et al.*, 1975; Flynn, 1971; Burton *et al.*, 1964). Octanol is ambiphilic and is capable of hydrogen bonding, as are phospholipids and proteins found in biological membranes. It is perhaps for these reasons that the octanol–water pair has become so widely used in quantitative structure–activity relationships (QSAR, see next section) in many areas of chemistry: physical, organic, biological, medicinal, quantum and computational (Hansch and Leo, 1995).

8 USEFULNESS OF THE OCTANOL–WATER PARTITION COEFFICIENT

8.1 QUANTITATIVE STRUCTURE–ACTIVITY RELATIONSHIPS

A great deal of interest in K_{ow} may be subsumed under the general rubric of quantitative structure–activity relationships (QSAR). Much of the credit for opening up this fertile field of investigation is due to Corwin Hansch and colleagues. Their early paper on the correlation of biological activity and chemical structure (Hansch and Fujita, 1964) led to an enormous amount of activity, in application, refinement and theory, by themselves and others. Significant work has been summarized in reviews and books (Leo *et al.*, 1971; Hansch *et al.*, 1989).

For present purposes, it suffices to note that a quantitative structure–activity relationship has the general form

$$BR = a + bB + cC + \ldots \tag{1.4}$$

where BR is biological response, $B, C \ldots$ are molecular properties and $a, b \ldots$ are fitting parameters. Hansch *et al.* (1989, 1995) provide detailed summaries of the very large number of individual studies resulting in Eq. (1.4). Here only a brief sketch is offered, in order to give the reader an idea of the extremely wide applicability of this equation.

The molecular properties $B, C \ldots$ usually appearing in the RHS of Eq. (1.4) are quite few: $\log K_{ow}$, $(\log K_{ow})^2$, π (hydrophobic parameter), σ (Hammett parameter), E_s (Taft steric parameter), molar refraction and molecular weight. A large fraction of QSARs were established with only $\log K_{ow}$ as independent variable:

$$BR = a + b \log K_{ow} \tag{1.5}$$

BR in Eq. (1.5) may be one of a large number of reactions of organisms or parts of organisms to the introduction of xenobiotic compounds (*in vivo* experiments). It may represent the interaction of a chemical in a biochemical reaction or process (*in vitro* experiments). BR commonly is $1/C$, where C is the molar concentration of chemical producing a standard indicated effect. This standard effect may be, for example:

LD_{50} (lethal dose for 50% of test organisms)
ED_{50} (effective dose for anaesthesia)
$\log k$, where k is a rate parameter (transport through membrane, permeation through skin, etc.)
inhibition of measurable biological processes

Hansch *et al.*, (1989, 1995) provide many examples of the applicability of Eq. (1.5) under the headings of non-specific toxicity, metabolism, central nervous system (CNS) agents, etc.

8.2 K_{ow} AND ENVIRONMENTAL 'PARTITION COEFFICIENT'

The quantitative structure–activity relationships represented by Eqs. (1.4) and (1.5) have obvious importance in the design or formulation of drugs and pharmaceuticals. K_{ow} has become equally important as a molecular parameter for describing the behaviour of xenobiotic chemicals in the environment. One such process is called *bioaccumulation* (Connell, 1990).

Bioaccumulation (bioconcentration, biomagnification) is the process by which a chemical accumulates in an organism to a higher concentration than is present in an external source. The defining equation is

$$[X]_{biota} = BCF\,[X]^n_{env} \tag{1.6}$$

where $[X]_{biota}$ is the concentration of a chemical in an organism, $[X]_{env}$ is the concentration in the environment (water, food, etc.) and BCF is the bioconcentration factor. The exponent n is usually equal to or close to 1. It has been found (Connell, 1990) that the BCF of a compound depends upon K_{ow} through a relation similar to Eq. (1.5), viz.,

$$\log BCF = a + b \log K_{ow} \tag{1.7}$$

For example, in pesticide bioconcentration by catfish (Ellgehausen *et al.*, 1980), $a = -1.71$ and $b = 0.83$.

In modelling the fate of organic pollutants in the biosphere, environmental chemists also use soil–water (partition) coefficients for compounds. For a large number of 'foreign' chemicals found in soil, the soil–water sorption coefficient K_{sw} was simply related to K_{ow} (Uchida and Kasai, 1980; Briggs, 1981):

$$\log K_{sw} = a + b \log K_{ow} \tag{1.8}$$

The same relationship was seen to be true for the sorption of organic compounds into wastewater solids (sludges) (Dobbs *et al.*, 1989) or marine sediments (Lara and Ernst, 1990). The partitioning of chemicals from water into cationic and anionic micelles can be described by a similar equation (Valsaraj and Thibodeaux, 1990).

K_{ow} has been used in a model to predict air–mammal tissue partition coefficients (Connell *et al.*, 1993); it is also an important parameter in a fugacity model for partitioning of chemicals among six environmental media (Mackay *et al.*, 1992–1995).

8.3 K_{ow} IN DATABASES

Good K_{ow} data exist for a large number of chemicals (Sangster, 1993) or can be calculated reliably from molecular structure (Chapter 5). Bioconcentration factors and other such environmental parameters are less studied and are usually more difficult or time-consuming to measure than K_{ow}. It is not surprising then that K_{ow} is one of a small number of properties included in electronic databases of environmental chemical information. Two examples can be cited. FATE is an online database of kinetic and equilibrium constants needed for assessing the fate of chemicals in the environment (Kollig and Kitchens, 1993). COMPUTOX is a database, in spreadsheet format, of standard toxicity data (McKinnon and Kaiser, 1993). Both databases contain K_{ow} as an important physiochemical property.

8.4 REASONS FOR THE SUCCESS OF K_{ow} IN BIOLOGICAL AND ENVIRONMENTAL APPLICATIONS

At first sight it may seem remarkable indeed that a single physiochemical property, K_{ow}, can be implicated in fundamental ways in such seemingly diverse areas as bioaccumulation and drug design. A little thought, however, reveals some basic interrelationships between physicochemical phenomena, on the one hand, and biological/biochemical and environmental processes on the other. K_{ow}, in thermodynamics, is what may be called a *free energy function* (like solubility and vapour pressure), and thus is directly concerned with the energetics of transfer of substances between phases.

The connection with biological activity was suggested by Hansch and Fujita (1964) and a simplified version of their diagram is shown in Figure 1.6. This shows a general reaction scheme for the interaction of a chemical compound with an organism (or part of an organism) to produce a biological response. The first step is thought of as a random walk process in which the molecule eventually arrives at a particular site in a cell from a dilute solution outside the cell. Once having arrived at this critical site, it undergoes a chemical reaction (or a series of reactions) to produce the measurable biological response. Step I is visualized as a relatively slow process (diffusion, permeation, etc.) which is highly dependent on the molecular structure of the compound concerned. Of course, however much the structure of the molecule may vary, its basic chemical properties remain the same so that it will react (bio)chemically in the same way. The rate-controlling step in this scheme is Step I, i.e. the 'partitioning' of the compound, going from one or more aqueous-like phases to organic-like phases. This step tends to be controlled by the energetics of partitioning, for which the partition coefficient is a good indicator.

The connection of K_{ow} with bioconcentration factor and other environmental partition constants follows from thermodynamic considerations.

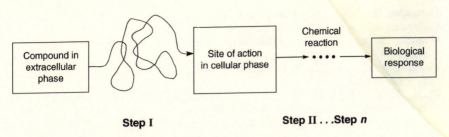

Figure 1.6 Simplified model for the interaction of a chemical compound and a biological system (Hansch and Fujita, 1964)

Hydrophobic chemicals tend to accumulate in organisms because they are metabolized only slowly and are effectively stored in tissue. The thermodynamic tendency to partition into organic phases (represented by K_{ow}) continues to operate, however, and the organism continues to absorb the chemicals as long as they are present in the environment. The connection of K_{ow} with sediment–water, soil–water and similar constants can be interpreted rather directly from thermodynamic considerations. In Chapter 2, it is shown that partition coefficients for other solvent–water pairs (solvent = alkanes, chloroform, diethyl ether, etc.) are directly related to those for the octanol–water pair. The physicochemical environment that a chemical 'sees' in sediments, sludges, soil or micelles will be more or less organic, 'oily', hydrocarbon-like, etc. In this way, the various environmental sorption constants can be seen as corresponding to a thermodynamic partitioning, approximating the model laboratory phenomenon.

REFERENCES

Ababi, V. and Popa, A. (1960) *An. Stiint. Univ. 'Al. I. Cuza' Iasi. Sect. I* **6**, 929–42.
Addison, C. C. (1945) *J. Chem. Soc.* 98–106.
Apelblat, A. (1983) *Ber. Bunsenges. Phys. Chem.* **87**, 2–5.
Beezer, A. E., Hunter, W. H. and Storey, D. E. (1980) *J. Pharm. Pharmacol.* **32**, 815–19.
Bernazzani, L., Cabani, S., Conti, G. and Mollica, V. (1995) *Thermochimica Acta* **269–270**, 361–9.
Berthelot, M. and Jungfleisch, E. (1872) *Ann. Chim. Phys. (4th Ser.)* **26**, 396–407
Berti, O., Cabani, S., Conti, G. and Mollica, V. (1986) *J. Chem. Soc. Faraday Trans. I* **82**, 2547–56.
Brodin, A., Sandin, B. and Faijerson, B. (1976) *Acta Pharm. Suec.* **13**, 331–52.
Briggs, G. G. (1981) *J. Agric. Food Chem.* **29**, 1050–9.
Burton, D. E., Clarke, K. and Gray, G. W. (1964) *J. Chem. Soc.* 1314–18.
Butler, J. A. V., Thomson, D. W. and Maclennan, W. H. (1933) *J. Chem. Soc.* 674–86.
Collander, R. (1949) *Acta Chem. Scand.* **3**, 717–47.
Collander, R. (1950) *Acta Chem. Scand.* **4**, 1085–98.

Collander, R. (1951) *Acta Chem. Scand.* **5**, 774–80.

Collander, R. (1954) *Physiol. Plantarum* **7**, 420–45.

Connell, D. W. (1990) *Bioaccumulation of Xenobiotic Compounds*, CRC Press, Boca Raton.

Connell, D. W., Braddock, R. D. and Mani, S. V. (1993) *Sci. Total Environ.* **Suppl. Part 2**, 1383–96 (1993).

Craig, L. C. and Craig, D. (1950) in *Technique of Organic Chemistry*, ed. A. Weissberger, Interscience, New York, Chap IV.

Crittenden, E. D. and Hixon, A. N. (1954) *Ind. Eng. Chem.* **46**, 265–8.

Dallas, A. J. and Carr, P. W. (1992) *J. Chem. Soc. Perkin Trans. 2*, 2155–61.

Dallos, A. and Liszi, J. (1995) *J. Chem. Thermodyn.* **27**, 447–8.

Dearden, J. C. and Bresnen, G. M. (1988) *Quant. Struct.-Act. Relat.* **7**, 133–44.

DeBolt, S. E. and Kollman, P. A. (1995) *J. Am. Chem. Soc.* **117**, 5316–40.

Dobbs, R. A., Wang, L. and Govind, R. (1989) *Environ. Sci. Technol.* **23**, 1092–7.

Ellgehausen, H., Guth, J. A. and Esser, H. O. (1980) *Ecotoxicol. Environ. Saf.* **4**, 134–57.

von Erichsen, L. (1952) *Brennstoff-Chem.* **33**, 166–72.

Flynn, G. L. (1971) *J. Pharm. Sci.* **60**, 345–53.

Francis, A. W. (1961) Critical Solution Temperatures, *Advances in Chemistry Series No. 31*, American Chemical Society, Washington.

Frank, H. S. (1972) 'Structural Models', Chap. 14 in *Water, a Comprehensive Treatise*, Vol. 1, ed. F. Franks, Plenum Press, New York.

Fujita, T., Iwasa, J. and Hansch, C. (1964) *J. Am. Chem. Soc.* **86**, 5175–80.

Hansch, C. and Fujita, T. (1964) *J. Am. Chem. Soc.* **86**, 1616–26.

Hansch, C., Kim, D., Leo, A., Novellino, E., Silipo, C. and Vittoria, A. (1989) *Crit. Rev. Toxicol.* **19**, 185–226.

Hansch, C. and Leo, A. (1995) *Exploring QSAR: Fundamentals and Applications in Chemistry and Biology*, American Chemical Society, Washington.

Hefter, G. T. (1984) *Solubility Data Ser.* **15**, 364–81.

James, M. J. (1983) Ph. D. thesis, University of Nottingham.

Kinoshita, K. Ishikawa, H. and Shinoda, K. (1958) *Bull. Chem. Soc. Jpn.* **31**, 1081–2.

Kollig, H. P. and Kitchens, B. J. (1993) *J. Chem. Inf. Comput. Sci.* **33**, 131–4.

Kristl, A. and Vesnaver, G. (1995) *J. Chem. Soc. Faraday Trans.* **91**, 995–8.

Lara, R. and Ernst, W. (1990) *Environ. Technol.* **11**, 83–92.

Lavrova, O. A. and Lesteva, T. M. (1976) *Zh. Fiz. Khim.* **50**, 1617.

Lebedinskaya, N. A., Kushner, T. M. and Ivanskaya, L. N. (1984) *Khim. Prom-st (Moscow)* 4, 206–7.

Leo, A., Hansch, C. and Elkins, D. (1971) *Chem. Rev.* **71**, 525–53.

Lide, D. R., ed. (1992) *Handbook of Chemistry and Physics*, 73rd edition, CRC Press, Boca Raton.

Liew, K. Y., Seng, C. E. and Ng, B. H. (1992) *J. Solution Chem.* **21**, 1177–83.

Mackay, D., Shiu, W. Y. and Ma, K. C. (1992–1995) *Illustrated Handbook of Physical-Chemical Properties and Environmental Fate for Organic Chemicals*, 4 vols., Lewis Publishers, Chelsea.

Marcus, Y. (1990) *J. Solution Chem.* **19**, 507–17.

McBain, J. W. and Richards, P. H. (1946) *Ind. Eng. Chem.* **38**, 642–6.

McKinnon, M. B. and Kaiser, K. L. E. (1993) *Chemosphere* **27**, 1159–69.

Meyer, M. (1899) *Exp. Pathol. Pharmakol.* **42**, 109–18.

Nernst, W. (1891) *Z. Phys. Chem.* **8**, 110–39.

Overton, C. E. (1901) *Studien über die Narkose, zugleich ein Beitrag zur allgemeinen Pharmakologie*, G Fischer, Jena.

Overton, C. E. (1991) *Studies on Narcosis*, ed. R. L. Lipnick, Chapman and Hall/Wood Library-Museum of Anaesthesiology, London.

Rao, K. S., Rao, M. V. R. and Rao, C. V. (1961) *J. Sci. Ind. Res. B* **20B**, 283–6.

Rekker, R. F. and Mannhold, R. (1992) *Calculation of Drug Lipophilicity*, VCH Verlagsgesellschaft, Weinheim.

Sangster, J. (1993) *LOGKOW — a Databank of Evaluated Octanol–Water Partition Coefficients*, Sangster Research Laboratories, Montreal.

Shinoda, K., Yamanska, T. and Kinoshita, K. (1959) *J. Phys. Chem.* **63**, 648–50.

Smith, R. N., Hansch, C. and Ames, M. M. (1975) *J. Pharm. Sci.* **64**, 599–606.

Sobotka, H. and Glick, D. (1934) *J. Biol. Chem.* **105**, 199–219.

Sørensen, J. M. and Arlt, W. (1979) *Liquid–liquid Equilibrium Data Collection*, Chemistry Data Series Vol. V, Part I, Deutsches Gesellschaft für Chemisches Apparatewesen, Frankfurt/Main.

Stephenson, R., Stuart, J. and Tabak, M. (1984) *J. Chem. Eng. Data* **29**, 287–90.

Tokunaga, S., Manabe, M. and Koda, M. (1980) *Niihama Kogyo Semmon Gakko Kiyo, Rikagaku Hen* **16**, 96–101.

Uchida, M. and Kasai, T. (1980) *Nippon Nagaku Gakkaishi* **5**, 553–8.

Valsaraj, K. and Thibodeaux, L. J. (1990) *Sep. Sci. Technol.* **25**, 369–95.

Vochten, R. and Petre, G. (1973) *J. Colloid Interface Sci.* **42**, 320–7.

Zhuravleva, I. K., Zhuravlev, E. F. and Lomakina, N. G. (1977). *Zh. Fiz. Khim.* **51**, 1700–7.

CHAPTER 2

Thermodynamics and Extra-thermodynamics of Partitioning

1 REVIEW OF CHEMICAL THERMODYNAMIC PRINCIPLES

The reader is assumed to be familiar with the elements of classical chemical thermodynamics. Here only a sketch is given as a reminder of pertinent quantities and relations used as a basis for this chapter. Fuller accounts of chemical thermodynamics are available in texts (Klotz and Rosenberg, 1986; Ragone, 1995; Reid, 1990). More detailed exposition is given in the subsections later in this chapter. Advanced treatments may be found in Guggenheim (1986), Lewis and Randall (1961) and Rowlinson and Swinton (1982).

1.1 SYSTEM AND SURROUNDINGS; FIRST LAW

In thermodynamics, the word *system* refers to any portion of space or matter set aside for study. The choice of what to include in a system is entirely that of the investigator. Though this may sound rather arbitrary, in practice it poses little problem in the situations discussed in this chapter. The *surroundings* are simply the rest of the universe, apart from the system. System and surroundings are separated by a defining boundary. Where to place the boundary is again rather arbitrary and is at the discretion of the investigator. It occasionally entails considerable economic consequences, such as in industrial energy conservation (Boustead and Hancock, 1979; Chiogioji, 1979).

Energy is transferred between system and surroundings as heat (Q) and work (W). Both Q and W are forms of energy. The internal energy U of a system depends upon inherent properties of the materials of the system and on temperature and pressure. (Contributions of electric, magnetic and gravitational fields, etc., are ignored in this treatment.) The change in U due to transfer of energy across the system boundary is given by

$$\Delta U = Q + W \qquad (2.1)$$

in which the convention is adopted that heat entering the system and work being done on the system are both positive quantities, both increasing the internal energy of the system.

Eq. (2.1) is another way of stating the principle of conservation of energy; it is understood that the mass of the system remains unchanged in the process. The treatment here is further simplified by assuming that there are no changes in kinetic and potential energy of the system as a whole during these changes.

As thus defined, the internal energy is a *state function*, which means that when U changes between two equilibrium state 1 and 2, the difference

$$\Delta U = U_2 - U_1 \tag{2.2}$$

is determined solely by the beginning and final states of the system and is independent of the thermodynamic path taken by the system to get from one state to the other.

The work W involved in thermodynamic processes may be of various forms: expansion or contraction of a gas, an electrochemical reaction, mechanical work, etc. In chemical thermodynamics it is useful to consider the first mentioned type, called *pressure–volume* (*P–V*) work. The quantity H, defined by

$$H \equiv U + PV \tag{2.3}$$

is called *enthalpy*. Since the PV contribution in Eq. (2.3) is by definition always evaluated under reversible conditions, H is a state property. H will depend upon temperature and pressure, as well as composition. For present purposes, the temperature dependence of H is important. Since most thermodynamic measurements of interest here are performed at constant pressure, the quantity C_p

$$C_p \equiv (\partial H \backslash \partial T)_p \tag{2.4}$$

is called the *heat capacity* (strictly speaking, energy capacity) at constant pressure.

1.2 IDEAL AND REAL GASES

An ideal gas is one which obeys the equation of state

$$PV = nRT \tag{2.5}$$

for n moles of gas. An ideal gas is, of course, a limiting case and its non-ideality may be expressed by a compressibility factor Z

$$PV = nZRT \tag{2.6}$$

For ordinary temperatures and pressures, Z for most gases and vapours is very close to unity. It can be estimated, with adequate accuracy, from a knowledge of critical constants, for both single gases and mixtures (Reid *et al.*, 1987).

1.3 SECOND LAW AND ENTROPY

The First Law, Eq. (2.1), is a statement about the conservation of energy in energy transfer with respect to heat and work. Eq. (2.2) is a reminder that the energy difference between two states is independent of the path taken from state 1 to state 2. The First Law, however, says nothing about whether or not the system will spontaneously undergo the change from state 1 to state 2.

In order to make thermodynamic statements about the spontaneity of a process, it is necessary to have a new state function, viz., *entropy*, S. The concept of entropy developed from some quite practical problems in the 19th century, namely, the study of devices to convert thermal energy (heat) into mechanical energy (work). The change in entropy is

$$\Delta S = S_2 - S_1 = Q_r/T \tag{2.7}$$

where Q_r is evaluated for the process operating in a reversible isothermal manner and T is the absolute temperature at which the change takes place. In contradistinction to U, which is conserved in a process (spontaneous or not), S is *not* conserved in a spontaneous process. For example, if heat flows from a hot reservoir (T_1) to a cooler reservoir (T_2), the changes in state functions for an isolated system are

$$\Delta U = 0 \tag{2.8}$$

$$\Delta S = Q(T_1 - T_2)/T_1 T_2 > 0 \tag{2.9}$$

1.4 PROPERTY RELATIONS

The thermodynamic processes of most interest here are those carried out at constant temperature and pressure. Under these conditions, another state function, called the *Gibbs free energy G*

$$G \equiv H - TS \tag{2.10}$$

serves, among other things, as a criterion for spontaneity (or equilibrium). Thus, for a system undergoing a change between two states, the process 1 → 2 will be spontaneous if

$$\Delta G = G_2 - G_1 < 0 \tag{2.11}$$

and if it happens that

$$\Delta G = 0 \tag{2.12}$$

then $G_2 = G_1$ and the system is said to be in equilibrium with respect to imposed conditions of T and P.

As neither U, H nor G for a substance have absolute values at given T and P, the 'zero point' for these quantities is arbitrary to some extent. It has been found convenient to define *reference states of elements* for which enthalpy H and Gibbs energy G are both zero:

Gases: pure ideal substance at 1 bar and 25°C;
Liquids: pure substance at 1 bar and 25°C;
Solids: pure substance in most stable (crystalline) state at 1 bar and 25°C.

The enthalpy or Gibbs energy of a compound — say CO_2 — can then be defined by a *formation reaction*

$$C(s) + O_2(g) \rightarrow CO_2(g) \tag{2.13}$$

at the standard conditions 25°C and 1 bar. If X represents H or G, the quantity

$$\Delta X = X(CO_2, g) - X(C, s) - X(O_2, g) \tag{2.14}$$

is called the *standard enthalpy (Gibbs energy) of formation*. ΔX in Eq. (2.14) can be determined experimentally. In this way, the enthalpies and Gibbs energies of formation of a very large number of substances can be collected and made available in compilations such as the JANAF Tables (Chase *et al.*, 1985).

1.5 ACTIVITY

The principal interest in this chapter is the thermodynamics of solutions and equilibrium. For this purpose, the Gibbs energy G must be further refined with the introduction of the function *thermodynamic activity*. For an ideal gas

$$dG = RT\, d \ln P \tag{2.15}$$

Eq. (2.15) is written in differential form because, since there is no absolute value of G, only *changes* in G are of practical interest. For a real gas

$$dG = RT\, d \ln f \tag{2.16}$$

where f is the *fugacity* (a kind of 'corrected' pressure). Since the behaviour of a real gas approaches that of an ideal gas as the pressure approaches zero,

$$\lim_{P \to 0} (f/P) = 1 \tag{2.17}$$

By integrating Eq. (2.16), we obtain

$$\Delta G = G_2 - G_1 = RT \int_1^2 d \ln f = RT \ln(f_2/f_1) \tag{2.18}$$

which gives the difference in Gibbs energy of a gas between two pressures at a given temperature. If state 1 be identified with a defined *standard state*, then the quantity

$$a \equiv f/f^\circ \tag{2.19}$$

defines the *activity* of the gas. The standard state of a substance can be arbitrarily defined, but the following convention has been generally adopted:

Gases: the pure ideal gas at 1 bar and the temperature of interest;
Liquids: the pure liquid at 1 bar and the temperature of interest;
Solids: the most stable form of the pure solid at 1 bar and the temperature of interest.

The fugacity of a condensed phase (liquid or solid) at a given temperature is equal to the fugacity of the vapour in equilibrium with it. The activity of solids and liquids is defined also by Eq. (2.19). It follows from Eqs. (2.18) and (2.19) that the difference of Gibbs energy between a substance in a given state and the same substance in the standard state is

$$\Delta G = G - G^\circ = RT \ln a \qquad (2.20)$$

1.6 SOLUTIONS

Hitherto the thermodynamic functions H, G, f and a have been discussed with respect to pure substances. When substances are part of a solution, the same quantities are retained. Since the properties of a substance in solution will in general be different from those of the pure substance, the partial molal quantity $\bar{X}$ ($X = H$, G, V, etc.) is used instead:

$$\bar{X}_i = (\partial X/\partial n_i)_{T,P,n_j,n_k} \ldots \qquad (2.21)$$

where n signifies number of moles and i indicates the substance of interest and j, k . . . represent all the other substances present in the solution. The total value of X for a solution made of number of components is

$$X = n_i \bar{X}_i + n_j \bar{X}_j + n_k \bar{X}_k + \ldots \qquad (2.22)$$

For Gibbs energy, the quantity $\bar{G}$ is often given the simpler symbol μ.

1.7 IDEAL SOLUTIONS

Earlier, it was found convenient to contrast real and ideal gases; for similar reasons it is advantageous to define an *ideal solution*. For such a solution, at all temperatures and pressures, the fugacity of a component is the product of its mole fraction and its fugacity in the standard state:

$$f_i = x_i f_i^\circ \qquad (2.23)$$

It follows from Eqs. (2.19) and (2.23) that, in an ideal solution,

$$a_i = x_i \qquad (2.24)$$

Eq. (2.24) is a succinct way of stating Raoult's law. The choice of standard states for components in a solution leads one naturally to think of the formation of a solution as resulting from the mixing of two or more pure substances together. It may be derived from Eq. (2.23) and fundamental thermodynamic principles that, when two or more substances are mixed to form an ideal solution

$$\Delta_{mix}H = \Delta_{mix}V = 0 \tag{2.25}$$

$$\Delta_{mix}G = -RT(x_i \ln x_i + x_j \ln x_j + x_k \ln x_k + \ldots) = \Delta_{mix}S \tag{2.26}$$

where $\Delta_{mix}X$ represents the total change in the property X upon mixing.

1.8 ACTIVITY COEFFICIENT AND EXCESS PROPERTIES

In the previous section, the ideal solution was defined as a convenient reference for comparison. Eq. (2.24) represents the behaviour of an ideal solution. Evidently then, for a real solution, $a_i \neq x_i$, but rather

$$a_i = \gamma_i x_i \tag{2.27}$$

where γ_i is called the *activity coefficient* and simply reflects the non-ideality of the solution with respect to component i. It is useful to refer to the thermodynamic properties of a real solution over and above those of the corresponding ideal solution; these are just differences (ΔX) indicated by X^E and are called *excess properties*. It follows from what has gone before that, for an ideal solution,

$$G^E = H^E = V^E = \Delta_{mix}H = \Delta_{mix}V = 0 \tag{2.28}$$

where G^E is zero because the ideal value of $\Delta_{mix}G$ already includes the ideal entropy of mixing term (Eq. 2.26). For real solutions, of course, excess properties may be different from zero. In particular, for G^E, we have

$$G^E = RT(\ln \gamma_i + \ln \gamma_j + \ln \gamma_k + \ldots) \tag{2.29}$$

or

$$G^E = \mu_i^E + \mu_j^E + \mu_k^E + \ldots \tag{2.30}$$

1.9 DILUTE SOLUTIONS

The excess properties of real solutions are, by definition, non-zero. Dilute solutions constitute a class of real solutions for which certain properties assume simplified forms. Since dilute solutions are frequently encountered in practice, some pertinent relations are discussed here.

For simplicity, we consider a solution of two components 1 and 2, in which the solvent 1 is in large excess. It is an empirical finding that, in such dilute solutions,

$$f_2 = K_H X_2 \qquad (2.31)$$

that is, the fugacity (or pressure, when the pressure is low) of the solute above the solution is proportional to its mole fraction in the solution. This is a statement of Henry's law, and the proportionality constant K_H is called the Henry's constant for the solute in that particular solvent. It can be shown, from a purely thermodynamic argument, that if the solute in a solution obeys Henry's law (Eq. 2.31), the solvent simultaneously obeys Raoult's law (Eq. 2.23).

If Eq. (2.31) be divided by f_2°, the standard state fugacity of the solute than

$$f_2/f_2^\circ = a_2 = K_H x_2/f_2^\circ = k x_2 \qquad (2.32)$$

where k is a new constant. Comparing Eqs. (2.27) and (2.32), we see that

$$\gamma_2 = \gamma_2^\infty = k \qquad (2.33)$$

where the symbol ∞ indicates that the solute is in an infinitely dilute state. The quantity γ^∞ may be called a Henrian activity coefficient. It has been found, for real dilute solutions, that over a finite region of concentration, γ_2 is constant within experimental uncertainty. This is usually referred to as the *Henry's law region*.

The behaviour of solutions in the Henry's law region gives rise to a number of simple relations for colligative solution properties. Besides those already discussed in this section, there are well-known consequences for the lowering of the freezing point of a solvent (or elevation of the boiling point), van't Hoff's law of osmotic pressure, etc. Another consequence, of particular importance for the thermodynamics of partitioning, is Nernst's distribution law. This will be elaborated in this chapter.

1.10 TRANSFER FUNCTIONS

In the consideration of thermodynamic solution properties, use is often made of quantities known as *transfer functions*. This nomenclature has been adopted principally for reasons of convenience and introduces no new thermodynamic concepts. In the present context, it is useful in representing changes in thermodynamic properties of solutes in different solvents (sometimes, though not necessarily, at infinite dilution). For example, if one measures the partial molal volume of a solute at infinite dilution in two solvents A and B, one can speak of the change in partial molal volume of the solute in going from solvent A to solvent B, writing it as

$$\Delta_{tr} V^\infty(A \rightarrow B) = V^\infty(B) - V^\infty(A) \qquad (2.34)$$

In this case, the process is assumed to take place at constant temperature and pressure.

2 EXCESS PROPERTIES OF THE OCTANOL–WATER SYSTEM

Although octanol and water are commonly thought to be immiscible, in fact there is a region of miscibility (Chapter 1). It is appropriate here to summarize known thermodynamic data of this system.

The activity of water in saturated and unsaturated alkanol solutions was determined by Apelblat (1983, 1990) at 25°C. The excess Gibbs energies of the miscible regions are shown in Figure 2.1 for the partially miscible C_4 to C_8 1-alkanols (methanol, ethanol and 1-propanol are all completely miscible with water). The data in Figure 2.1 are from Apelblat (1990) for the C_5 to C_8 alkanols; for 1-butanol an average was taken of the results of Apelblat (1990) and excess Gibbs energies derived from the vapour–liquid equilibrium data of Butler *et al.*, (1993) and Lyzlova *et al.* (1979). These curves show the expected positive excess property, which increases with carbon number.

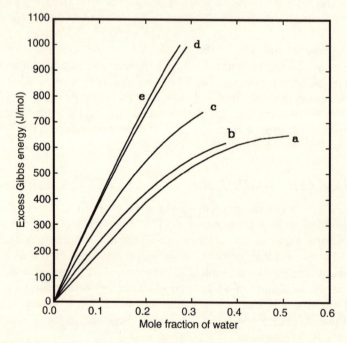

Figure 2.1 Excess Gibbs energies of l-alkanol/water systems at 25°C: a, butanol; b pentanol; c, hexanol; d, heptanol; e, octanol.

Table 2.1 Henrian activity coefficients of both components in 1-alkanol/water systems at 25°C

1-alkanol	γ^∞	
	1-alkanol in water (Abraham, 1984)	water in 1-alkanol[*]
Methanol	2.49	1.5
Ethanol	3.73	2.7
Propanol	14.2	4.9
Butanol	52.8	5.5
Pentanol	214	7.8[a]
Hexanol	856	9.7
Heptanol	3594	11[b]
Octanol	12534	13[b]

[*]mean values selected from Gmehling and Onken (1977) and Gmehling et al. (1981).
[a]interpolated.
[b]extrapolated.

For immiscible systems, the thermodynamic quantities at infinite dilution are of particular interest. Table 2.1 presents the Henrian activity coefficients of both components, at 25°C, for l-alkanol/water systems. The activity coefficients of the alkanols show a marked dependence on carbon number, as the higher members of the homologous series become increasingly insoluble. In contrast, the activity coefficient of water is relatively much less sensitive to the carbon number of the alkanol; evidently the alcoholic hydroxyl groups are more important than the hydrocarbon chains in determining the effective environment of a water molecule in alcohol solutions.

Of equal interest are the excess enthalpies at infinite dilution (heat of solution at infinite dilution). Table 2.2 summarizes recent reliable data. The corresponding value for water in octane is 34 kJ/mol (Nilsson, 1986a) and 35 kJ/mol for octane in water (Abraham, 1988), confirming the importance of the hydroxyl group in long-chain alkanols. The results for l-pentanol, l-hexanol and l-heptanol at temperatures others than 25°C (Hill and White, 1974; Pfeffer et al., 1995) suggest that $H^{E,\infty}$ for higher alcohols become positive at higher temperatures. This implies that the temperature dependence of solubility of long-chain alcohols in water, changes sign above 25°C. i.e., that the solubility–temperature relationship is not monotonic.

3 LOG K_{ow} AND SOLUBILITY

A method often used to derive K_{ow} for a particular compound is the use of an empirical correlation with water solubility. Such a possibility is attractive, since

Table 2.2 Enthalpy of solution at infinite dilution of both components in 1-alkanol/water systems at 25°C

	$\Delta_{sol}H^{\infty}$(kJ/mol)	
	1-alkanol in water Hallén et al. (1986)	water in 1-alkanol Nilsson (1986b)
Methanol	−7.30	−2.99
Ethanol	−10.16	−2.01
Propanol	−10.16	0.20
Butanol	−9.27	1.68
Pentanol	−7.96	2.66
Hexanol	−6.41	3.05
Heptanol	−4.88	3.41
Octanol	−3.37	3.31
Decanol	—	3.24

the solubility in water at ambient temperature is a well-documented property of many organic compounds. A correlation of this type was suggested by Hansch et al. (1968) and Chiou et al. (1977) and has since proved to be valuable in drug and pharmaceutical chemistry (Leahy, 1986) and environmental chemistry (Patil, 1991; Patil and Bora, 1994).

A number of correlations were proposed (Isnard and Lambert, 1989), summarized by Niimi (1991) and all were successful to some degree. The general relation is

$$\log K_{ow} = a \log S_w + b \tag{2.35}$$

where S_w is solubility in water (sometimes, though not invariably, in molarity) and a and b are empirical coefficients. As shown in this section, Eq. (2.35) has a thermodynamic basis (Mackay et al., 1980; Miller et al., 1984; Sangster, 1989; Tewari et al., 1982) but is not exact; the empirical coefficients in Eq. (2.35) depend somewhat on the type of compounds being considered. A thermodynamic derivation of Eq. (2.35) will show the source of this dependency. For reasons of simplicity of presentation, octanol and water are here considered to be completely immiscible; the effect of mutual solubility will be examined later in this chapter. It will simplify the thermodynamic treatment if volume fractions (ϕ) and volume fraction activity coefficients (f) are used; this choice follows from the definition of K_{ow} which is volume-based, rather than mole fraction based. Concentration (molarity) is indicated by c. Conversion between solute mole fraction and solute molarity may be made through the relation (valid for dilute solutions)

$$x = cV_{solv} \tag{2.36}$$

where V_{solv} is the molar volume of the solvent (dm^3/mol).* Liquid solutes will be considered first, as solid solutes require the introduction of complicating factors. If a liquid solute is distributed between water and octanol at equilibrium, we can write for the solute in each phase

$$\mu = \mu^\circ + RT \ln a \qquad (2.37)$$
$$= \mu^\circ + RT \ln (f\phi)$$

where μ° is the chemical potential of pure liquid solute. Thus at equilibrium $\mu^{oct} = \mu^w$ and so, from Eq. (2.38),

$$(f\phi)^{oct} = (f\phi)^w \qquad (2.39)$$

Rearranging Eq. (2.39),

$$f^w/f^{oct} = \phi^{oct}/\phi^w \qquad (2.40)$$

By definition, we have, in each phase

$$cV = \phi \qquad (2.41)$$

where V is the (partial) molal volume of solute in solution.† For dilute solutions of non-electrolytes in water or octanol, partial molal volume can be replaced by pure liquid molal volume without appreciable error. Substituting Eq. (2.41) in Eq. (2.40) we have

$$f^w/f^{oct} = c^{oct}/c^w = K_{ow} \qquad (2.42)$$

Eq. (2.42) shows that K_{ow} is equivalent to the ratio of the volume-based Henrian activity coefficients of the solute in the phases.

In order to derive the solubility relation Eq. (2.35) we must suppose that both water and octanol phases are saturated with solute; additionally, we require that the saturated solutions still be Henrian and that Eq. (2.42) applies. Since the solute in both phases is in equilibrium with pure liquid solute (assumed not to dissolve any solvent), $\mu = \mu^\circ$ in Eqs. (2.37) and (2.38) and hence

$$(f\phi)^{ss} = 1 \qquad (2.43)$$

---

*Here and hereafter, variables with no subscript are assumed to refer to the *solute*; variables with subscripts refer to the *solvent*, which may be indicated 'oct' or 'w'. Superscripts refer to the solution phase.†The ϕ in Eqs. (2.37)–(2.40) is, strictly speaking, a *formal* volume fraction (based on molar volumes of the pure components). The V in Eq. (2.41) is, however, necessarily a *partial* molal volume since its accompanying c (molarity) refers to one litre of *actual* (not formal) solution. Thus in Eq. (2.41) the RHS and LHS refer to different types of solution (formal/ideal and actual/real). For most organic liquids in water and octanol, however, Eq. (2.41) will introduce negligible error into the thermodynamic argument.

for each phase, where ss signifies a solution saturated with respect to the solute. Substituting Eq. (2.41) into Eq. (2.43) for the water phase,

$$(fcV)^{w,ss} = 1 \tag{2.44}$$

Thus, from Eqs. (2.42) and (2.44)

$$\log K_{ow} = \log \left(f^w / f^{oct} \right)^{ss} = \log[1/(cV)^{w,ss} f^{oct,ss}] \tag{2.45}$$

or

$$\log K_{ow} = -\log c^{w,ss} - \log \left(f^{oct,ss} V \right) \tag{2.46}$$

in which it has been assumed that partial molal volume of solute is equivalent to pure molal volume. Eq. (2.46) may be written in equivalent fashion on a mole fraction basis. The activity of the solute is the same, irrespective of units:

$$f\phi = \gamma x \tag{2.47}$$

which is valid for the solute in either liquid phase. By appropriate substitution of Eqs. (2.36) and (2.41) in Eq. (2.47),

$$f = \gamma V_{solv} / V \tag{2.48}$$

Substituting Eq. (2.48) in Eq. (2.46),

$$\log K_{ow} = -\log c^{w,ss} - \log \left(\gamma^{oct,ss} V_{oct,ss} \right) \tag{2.49}$$

Now $c^{w,ss}$ is precisely the solubility (molarity) of the solute in water and Eqs. (2.46) and (2.49) have the form of Eq. (2.35). The empirical predictive adequacy of Eqs. (2.35) and (2.49) will be discussed in Chapter 4.

For solutes which are solid at the K_{ow} measurement temperature, the thermodynamic treatment is similar. In this case Eq. (2.38) is still valid, but when a solid solute is in equilbrium with a saturated liquid solution, we have

$$\mu(s) = \mu(\text{solution}) \tag{2.50}$$

Then

$$\mu^o(s) + RT \ln a(s) = \mu^o(l) + RT \ln a(\text{solution}) \tag{2.51}$$

On the RHS of Eq. (2.51) the standard state is the pure liquid (here, a supercooled liquid). Rearranging,

$$\mu^o(l) - \mu^o(s) = RT \ln a(s) - RT \ln a(\text{solution}) \tag{2.52}$$

The quantity on the LHS of Eq. (2.52) is the Gibbs energy of fusion of the solute at temperature T. If the equilibrium solid phase is the pure solute (does not dissolve any solvent), $a(s) = 1$ and so

$$\Delta_{fus}G_T^o = -RT \ln a(\text{solution})$$
$$= -RT \ln(f\phi)^{ss} \tag{2.54}$$

as in Eq. (2.38). The ensuing thermodynamic treatment is similar to the case for a liquid solute, and the final relation is

$$\log K_{ow} = -\Delta_{fus}G_T^o/2.303RT - \log c^{w,ss} - \log(Vf^{oct,ss}) \tag{2.55}$$

which is to be compared to Eq. (2.46). There is a choice of ways of representing the first term on the RHS of Eq. (2.55). It may be written, from the relation $\Delta G = \Delta H - T\Delta S$

$$\Delta_{fus}G_T^o = \Delta_{fus}H^o - T\Delta_{fus}H^o/T_{fus} = \Delta_{fus}H^o(1 - T/T_{fus}) \tag{2.56}$$

where the heat of fusion is that at T_{fus}. This may be called a 'zero order' approximation. It is equivalent to the assumption that there is no change in heat capacity through the melting point, i.e., $\Delta_{fus}C_p^o = C_p^o(l) - C_p^o(s) = 0$. If $\Delta_{fus}C_p^o \neq 0$, then is follows from thermodynamics that

$$\Delta_{fus}G_T^o = \Delta_{fus}H^o - T_{fus}\Delta_{fus}C_p^o + T[\Delta_{fus}C_p^o \ln(T_{fus}/T) - \Delta_{fus}S^o] \tag{2.57}$$

where the heat and entropy of fusion are those at T_{fus}.

Eq. (2.46) gives K_{ow} in terms of solubility in water; a similar derivation gives the corresponding relation with solubility in octanol:

$$\log K_{ow} = \log c^{oct,ss} + \log(Vf^{w,ss}) \tag{2.58}$$

The solubility of environmental pollutants (Anliker and Moser, 1987; Niimi, 1991; Pinsuwan et al., 1995) and pharmaceuticals (Dagorn et al., 1985; Hatanaka et al., 1990; Kosanovic et al., 1988; Yalkowsky et al., 1983) have been measured in octanol.

The symmetrical relations, Eqs. (2.46) and (2.58), suggest that K_{ow} may be closely related to the octanol–water solubility ratio (Yalkowsky et al., 1983). We apply Eq. (2.42) to solute–saturation conditions:

$$K_{ow} = c^{oct,ss}/c^{w,ss} \approx f^w/f^{oct} \tag{2.59}$$

where it has been assumed that the activity coefficients remain Henrian up to saturation. Yalkowsky et al. (1983) tested Eq. (2.59) with solubility data of 36 solid compounds in neat solvents and found reasonable, though not exact, correlation.

4 TRANSFER OF SOLUTE AMONG PURE, GASEOUS AND SOLUTION PHASES

The thermodynamics of the process in which a pure solute enters a solvent to form a solution has been widely studied. Much useful information concerning solute–solvent interactions, among others, can be established by this route. A commonly used scheme relating the processes is shown in Figure 2.2 (Abraham, 1984). The process labelled 'solution' represents the ordinary mixing process from which excess properties are derived. That labelled 'solvation' is often used in those cases in which solute–solute interactions are to be bypassed. As Abraham (1984) shows, the choice of concentration scales (and hence of standard state) must be explicitly known before any data can be interpreted.

In Section 1, the properties of dilute solutions and transfer functions were briefly mentioned. The obvious usefulness of knowledge of the Gibbs energy of liquid–liquid distribution (partitioning) is well-attested; such knowledge should be complemented by equally important knowledge of the enthalpy and entropy of partitioning (Tomlinson *et al.*, 1986). Such complementary data, with respect to the octanol–water pair, have only recently been measured in a rigorous and systematic manner.

The partition coefficient K_{ow} is defined in terms of molarities of the solute. For thermodynamic treatment, a partition coefficient based on mole fraction is more familiar. Such a ratio, here designated as K_{owx}, may be obtained by substituting Eq. (2.36) in the definition. Thus,

$$K_{owx} = K_{ow} V_{oct} / V_w \qquad (2.60)$$

where the Vs are the molar volumes of the pure solvents. The Gibbs energy of transfer of solute from water to octanol can then be defined by

$$\Delta_{tr} G(w \rightarrow oct) = -RT \ln K_{owx} \qquad (2.61)$$

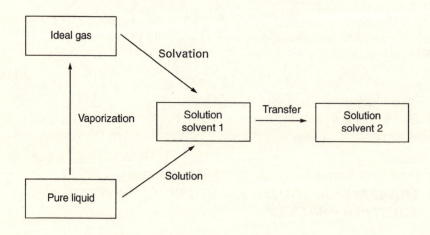

Figure 2.2 Transfer of a solute among media

where 'tr' indicates a transfer function. Other transfer functions follow from Eq. (2.61) in the usual manner:

$$\Delta_{tr}G = \Delta_{tr}H - T\Delta_{tr}S \tag{2.62}$$

$$[\partial(\Delta_{tr}G)/\partial T]_p = -\Delta_{tr}S \tag{2.63}$$

$$[\partial(\Delta_{tr}G/T)/\partial(1/T)]_p = \Delta_{tr}H \tag{2.64}$$

Eqs. (2.62–2.64) indicate that the temperature dependence of K_{ow} can be related to the enthalpy and entropy of transfer. Now K_{ow} has been measured as a function of temperature for a number of compounds (see next section) and hence the data could be a source of knowledge of $\Delta_{tr}H$ and $\Delta_{tr}S$ for these compounds. The temperature dependence of K_{ow} is sometimes written as

$$d(\log K_{ow})/dT = a/T + b \tag{2.65}$$

(a and b are fitting parameters), which relation is sometimes called a 'van't Hoff' plot. An approximate value for $\Delta_{tr}H$ can be derived from a knowledge of the parameters of Eq. (2.65), but in practice K_{ow} is quite insensitive to temperature and the corresponding uncertainty in $\Delta_{tr}H$ is large. Furthermore, the mutual solubilities of water and octanol are temperature dependent, and hence K_{ow} (as measured) represents a different transfer process at every temperature. The quantity $\Delta_{tr}H$ can be measured more directly and more accurately, however. One can, for instance, perform calorimetric heat-of-mixing experiments using a solute with each of two solvents (water and octanol). If this is done in the dilute region, then the limiting heats of mixing, yield the transfer functions:

$$\Delta_{tr}H(w \rightarrow oct) = H^{E,\infty}(oct) - H^{E,\infty}(w) \tag{2.66}$$

where the quantities on the RHS are excess (partial) enthalpies at infinite dilution. Another method, recently invented, makes use of an *isoperibol calorimeter* (Riebesehl and Tomlinson, 1983). In this instrument, immiscible solvent streams are brought into direct contact, solute is transferred across the liquid–liquid boundary and the heat effect is measured. Some enthalpies of transfer, measured calorimetrically, are presented in Table 2.3.

5 VARIATION OF K_{ow} WITH TEMPERATURE

It has been found that log K_{ow}, as measured for a given compound, is rather insensitive to temperature (Leo *et al.*, 1971). Table 2.4 summarizes experimental measurements of the temperature dependence of log K_{ow} for a variety of compounds. As stated in Section 4, a van't Hoff estimate of the

Table 2.3. Calorimetric measurements of the enthalpy of transfer, from water to octanol, of some compounds at 25°C

Compound	Method*	$\Delta_{tr}H$ (kJ/mol)	Reference
Methanol	IP	6.63	Riebesehl and Tomlinson (1986)
1-Butanol	IP	10.36	Riebesehl and Tomlinson (1986)
	HS	9.74	Cabani et al. (1991)
1-Heptanol	IP	5.13	Riebesehl and Tomlinson (1986)
Hexane	HS	0.81	Cabani et al. (1991)
Octane	HS	−0.45	Cabani et al. (1991)
1,4-Dioxane	HS	18.92	Cabani et al. (1990)
Diethylether	HS	22.15	Cabani et al. (1991)
Morpholine	HS	22.44	Cabani et al. (1990)
Cyclohexane	HS	1.65	Cabani et al. (1991)
Acetone	HS	18.52	Cabani et al. (1991)
Ethylenediamine	HS	15.66	Cabani et al. (1990)
Triethylamine	HS	26.10	Berti et al., (1987)
Dipropylamine	HS	21.48	Berti et al. (1987)
N-methylpiperazine	HS	32.39	Cabani et al. (1990)
N-methylpiperidine	HS	22.53	Cabani et al. (1991)
2-Methoxyphenol	HS-MSS	−1.45	Beezer et al. (1987)

*IP=isoperibol calorimetry (mutually saturated solvents); HS=heat of solution in separate pure solvents; HS-MSS=heat of solution in separate mutually saturated solvents.

enthalpy of transfer may be obtained from the relation, Eq. (2.65). There are few instances in which such van't Hoff enthalpies can be compared directly to calorimetrically derived data. In the case of some resorcinol monoethers, Beezer *et al.* (1980, 1987) made such measurements. The results are compared in Table 2.5. In both reports, the transfer refers to mutually saturated solvents. Approximate equivalence is evident.

Transfer thermodynamics has sometimes been discussed from the point of view of 'enthalpy-entropy compensation' (Tomlinson, 1983). This phenomenon manifests itself most clearly as a linear relationship between $\Delta_{tr}G$ and $\Delta_{tr}H$ for a homologous series of compounds in a single solvent pair. Some examples are shown in Figure 2.3. This relationship is valid for homologous series involving linear alkyl chains, and for calorimetric enthalpies of transfer, *not* van't Hoff enthalpies (Kinkel *et al.*, 1981). The linear relationship shown in Figure 2.3 is not displayed by unrelated organic compounds (Riebesehl *et al.*, 1984). Although the enthalpy–entropy compensation effect, when it occurs, is striking, its ultimate significance is unclear. It may simply reflect the CH_2 increment to the Gibbs energy and enthalpy of transfer.

Transfer property regularities in chemically similar and homologous series of compounds are more evident with respect to the solvation process. When the solvation process (Figure 2.1) refers to water as solvent, the process is usually called *hydration*; for any other solvent, *solvation* is used. Thus ΔG and/or ΔH

Table 2.4 Temperature coefficient of $\log k_{ow}$ at 25°C of various compounds, by direct measurement

Compound(s)	Temperature range (°C)	d($\log K_{ow}$)/dT $\times 1000$, K^{-1}	Reference
Propl benzene	10–35	−1.9	DeVoe et al. (1981)
Phenol	10–60	3.4	Korenman and Udalova (1974)
Phenol	20–50	−4.9	Rogers and Wong (1980)
Monosubstituted phenols	10–60	1.2 to 8.2	Korenman and Udalova (1974)
Mono- and disubstituted phenols	20–50	−13. to +9.5	Rogers and Wong (1980)
Resorcinol monoethers	15–35	−4.7 to −3.0	Beezer et al. (1980)
Chlorobenzenes	13–33	−18. to −7.	Opperhuizen et al. (1988)
Mono- and disubstituted benzenes	20, 60	−6.4 to +9.6	Kramer and Henze (1990)
Ephedrine	15–40	8.2	Brodin et al. (1976)
Methamphetamine	15–40	9.7	Brodin et al. (1976)
Methyl nicotinate	5–25	7.2	Guy and Honda (1984)
Pyridyl alkanamides	20–40	2.2 to 6.8	Repond et al. (1987)
Substituted phenazines	20–55	5.9 to 32.	Quigley et al. (1990)
End-protected tripeptides	2–65	6.0 to 19.	Kim and Szoka (1992)
Anti-inflammatory drugs	15–53	−18. to zero	Betageri et al. (1996)

Table 2.5 Enthalpies of transfer of resorcinol monoethers [(3-RO)C_6H_4OH] from water to octanol at 25°C, by calorimetric and van't Hoff methods (Beezer et al., 1980, 1987)

R	$-\Delta_{tr}H$(kJ/mol)	
	van't Hoff	calorimetric
Methyl	7.91	8.03
Ethyl	7.20	6.95
Propyl	5.92	6.96
Butyl	5.22	(7.18)*

*Predicted value

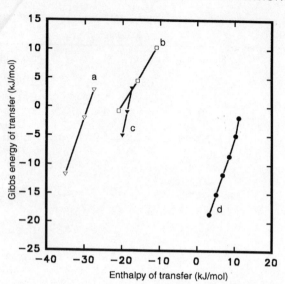

Figure 2.3 Enthalpy-entropy compensation for homologous series in different solvent pairs: a, 1-alkanols, water/2,2,4-trimethylpentane (Riebesehl *et al.*, 1984); b, 1-alkanols, water/chloroform (Riebesehl *et al.*, 1985); c, 2-alkanones, water/2,2,4-trimethylpentane (Riebesehl *et al.*, 1984); d, 1-alkanols, water/octanol (Riebesehl and Tomlinson, 1986).

for the process (gas→aqueous solution) are linearly related to the carbon number in homologous series (Abraham, 1984). Such linear relationships have also been found for other water partners, such as cyclohexane, octane, 2,2,4-trimethylpentane and octanol (Bernazzani *et al.*, 1995; Riebesehl and Tomlinson, 1984 and 1986).

6 IMPLICATIONS OF MUTUALLY SATURATED SOLVENTS

Hitherto in this chapter, the simplifying assumption has been made—wherever possible—that octanol and water are totally immiscible. As shown in Chapter 1, however, this is not strictly true. The saturation mole fractions are, at 25°C: water in octanol, 0.275; octanol in water, 7.5×10^{-5}. On a mole fraction basis, therefore, the solubility of octanol in water is several orders of magnitude smaller than that of water in octanol.

6.1 EFFECTS ON DENSITY

The difference in densities of pure water and octanol-saturated water at 25°C is virtually undetectable (Dallos and Liszi, 1995). This is consistent with the very low solubility of octanol in water. For water in octanol, however, there are measurable differences. The density of water-saturated octanol at 25°C is 0.8279 g/cm³ (Bowman and Sans, 1983) or 0.8288 g/cm³ (Dallos and Liszi, 1995). Hence the molar volume of pure octanol changes from 160.8 cm³/mol to 124.2±2.4 cm³/mol for water-saturated octanol.

6.2 EFFECTS ON SOLUBILITY

From a simple-minded point of view, it would seem that the presence of octanol in water would increase the solubility of lipophilic compounds and decrease the solubility of hydrophilic compounds; the presence of water on octanol would do the opposite. Table 2.6 and 2.7 present data on measured solubilities in neat and mutually saturated solvents. It appears that the presence of the 'other' solvent has little effect on the solubility of hydrophobic chemicals (Miller $et\ al.$, 1985), but has predictable effects on the solubility of hydrophilic compounds (Kristl, 1996; Kristl and Vesnaver, 1995).

Yalkowsky $et\ al.$ (1983) measured the solubility ratio $c^{oct,ss}/c^{w,ss}$ of four compounds, using both pure and mutually saturated solvents. The effect of mutual saturation was to increase the ratio in every case.

Table 2.6 Solubility at 25°C of various compounds in water and in octanol-saturated water (c=molarity)

Compound	Solubility		Reference
	log c (water)	log c (octanol-saturated water)	
Naphthalene	−3.61	−3.65	Miller $et\ al.$ (1985)
Biphenyl	−4.31	−4.35	Miller $et\ al.$ (1985)
1-Methylfluorene	−5.22	−5.24	Miller $et\ al.$ (1985)
Anthracene	−6.59	−6.76	Miller $et\ al.$ (1985)
Chrysene	−8.28	−7.54	Miller $et\ al.$ (1985)
2,3-Benzofluorene	−6.73	−6.77	Miller $et\ al.$ (1985)
4-PCB	−5.21	−5.23	Miller $et\ al.$ (1985)
4,4'-PCB	−6.56	−6.81	Miller $et\ al.$ (1985)
m-alkoxyphenols			
R=methyl	−0.51	−0.62	Beezer $et\ al.$ (1983)
R=propyl	−1.59	−1.73	Beezer $et\ al.$ (1983)
Acyclovir	−2.15	−2.16	Kristl and Vesnaver (1995)
Deoxyacyclovir	−1.08	−1.08	Kristl and Vesnaver (1995)
Diacetyldeoxy-acyclovir	−0.73	−0.74	Kristl and Vesnaver (1995)

Table 2.7 Solubility at 24±1°C of various compounds in octanol and water-saturated octanol (c=molarity)

Compound	Solubility		Reference
	log c (octanol)	log c (water-saturated) octanol	
Naphthalene	−0.018	−0.006	Miller *et al.* (1985)
Biphenyl	−0.160	−0.159	Miller *et al.* (1985)
1-Methylfluorene	−0.556	−0.655	Miller *et al.* (1985)
Anthracene	−1.93	−1.88	Miller *et al.* (1985)
Chrysene	−2.70	−2.73	Miller *et al.* (1985)
2,3-Benzofluorene	−1.75	−1.75	Miller *et al.* (1985)
4-PCB	−0.216	−0.233	Miller *et al.* (1985)
4,4′-PCB	−1.15	−1.18	Miller *et al.* (1985)
Acyclovir	−3.96	−3.77	(Kristl and Vesnaver (1995)
Deoxyacyclovir	−3.04	−2.34	Kristl and Vesnaver (1995)
Diacetyldeoxy-acyclovir	−2.60	−1.86	Kristl and Vesnaver (1995)

6.3 EFFECTS ON THERMODYNAMIC PROPERTIES

The presence of small amounts of the other solvent in octanol and water, raised questions about their thermodynamic significance. It was found that both enthalpy and Gibbs energy of transfer might be significantly altered by the presence of the other solvent (Berti *et al.*, 1986). The effects could be positive or negative or zero, depending upon circumstances. Tables 2.8 and 2.9 present some data on measured enthalpies and Gibbs energies illustrating this finding. In converting various volume-based quantities in Table 2.8 to mole fraction basis, the following molar volumes were used (cm^3/mol): water, 17.96; pure octanol, 160.8; water-saturated octanol, 124.2.

It should be pointed out, in assessing the significance of the experimental measurements and conclusions made from them, that due attention is to be paid to experimental bias and imprecision: some apparent effects might well be within the limits of experimental uncertainty. Care must also be taken in using results obtained by different methods. This is particularly true in the case of Gibbs energies of transfer (Table 2.8). Although different authors may occasionally report significant results for a single compound in a particular solvent, data in Table 2.8 imply that the octanol–water partition coefficient is not significantly affected by the presence of the 'other' solvent in the phases (Schantz and Martire, 1987; Wasik *et al.*, 1981). The existence of apparent exceptions like acetone, 2-butanone, butylamine and dipropylamine (Bernazzani *et al.*, 1995) might be due to experimental error.

Table 2.8a Gibbs energies of transfer (kJ/mol) from water to octanol at 25°C, for neat and mutually saturated solvents. Solute standard state: unit mole fraction

Compound	Neat solvents			Mutually saturated solvents		
	A*	B	C	A	B	C
Pentane		−26.0	26.2		−26.1	−25.5
Hexane	−29.8	−29.2	−29.4	−27.7	−28.9	−28.3
Heptane	−33.4	−32.4	−32.6	−32.0	−32.0	−31.4
Octane	−36.3	−35.4	−35.6	−35.0	−35.0	−34.4
1-Hexene		−24.8	−25.3		−24.8	−24.6
1-Heptene		−28.6			−28.2	
1-Octene		−32.4	−32.6		−31.5	−32.7
1-Nonene		−35.7	−35.9		−34.8	−35.3
Benzene		−16.8	−17.0		−16.9	−17.0
Toluene		−20.5	−20.6		−20.6	−19.9
Ethylbenzene			−23.4			−22.7
o-Xylene			−23.2			−22.7
m-Xylene			−24.0			−23.1
Propylbenzene		−26.6	−26.7		−26.5	−25.9
Butylbenzene		−30.2			−29.9	
Cyclohexane	−26.2			−25.1		
1-Bromobutane		−20.5	−21.4		−21.1	−20.5
1-Bromopentane		−24.4			−24.7	
1-Bromohexane		−27.1			−27.1	
1-Bromoheptane		−30.8			−30.3	
1-Bromooctane		−34.5			−33.3	
Diethylether	−11.10			−10.5		
Tetrahydrofuran	−7.10			−8.06		
Tetrahydropyran	−10.4			−9.09		
Acetone	−2.24			−4.06		
2-Butanone	−5.90			−7.09		
Diethylketone	−9.15		−9.74	−11.1		−10.4
2-Hexanone	−12.1			−13.3		
2-Heptanone			−17.3			−16.1
1-Aminobutane	−8.06			−10.3		
Dipropylamine	−10.6			−15.1		
Chlorobenzene			−21.6			−20.8
Ethylacetate			−8.45			−8.68
Propylacetate			−11.7			−11.9
Butylacetate			−15.1			−15.2

*A=Bernazzani *et al.* (1995); B=Schantz and Martire (1987); C=Wasik *et al.* (1981)

6.4 EFFECTS ON LOG K_{ow}–WATER SOLUBILITY CORRELATION

In section 3, a thermodynamic relation was derived between log K_{ow} and log $c^{w,ss}$, the molar solubility in water, on the assumption of totally immiscible solvents. Now octanol and water are in fact mutually miscible to some extent, and the derivation represented by Eqs. (2.37)–(2.58) may be repeated for

Table 2.8b Gibbs energies of transfer (kJ/mol) of alcohols at 25°C from water to octanol, for neat (mutually saturated solvents). Solute standard state: unit mole fraction

Compound	A*	B	C	D
Methanol		−0.41 (−0.59)		
Ethanol		−2.71 (−2.85)		
1-Propanol		−6.06 (−6.09)		
2-Propanol		−4.75 (−4.79)		
1-Butanol	−7.95 (−9.72)	−9.45 (−9.38)	−9.92 (−9.92)	−9.80 (−9.31)
2-Butanol	−7.66 (−9.49)			
Isobutanol		−9.18 (−9.00)		
tert-Butanol	−5.55 (−7.49)	−6.47 (−6.41)		
1-Pentanol		−12.9 (−12.8)	−14.3 (−14.2)	
Isopentanol		−12.2 (−12.0)		
1-Hexanol		−16.3 (−16.3)	−17.1 (−17.0)	
Cyclohexanol		−11.7 (−11.7)		
1-Heptanol			−20.1 (−20.1)	

*A=Bernazzani *et al.* (1995); B=Dallas and Carr (1992); C=Schantz and Martire (1987); D=Wasik *et al.* (1981)

mutually saturated solvents. A more realistic derivation, taking mutual solubility into account, was reported by Chiou *et al.* (1982). Mutually saturated solvents are distinguished here by a superscript asterisk (Bernazzani *et al.*, 1995; Bowman and Sans, 1983; Cabani *et al.*, 1991; Chiou *et al.*, 1982).

The key relationships for mutually saturated solvents now follow. Eq. (2.42) becomes

$$K_{ow} = c^{*oct}/c^{*w} = f^{*w}/f^{*oct} \tag{2.67}$$

The water solubility in the usual correlation relation, Eq. (2.35), remains solubility in pure water, not the saturated solvent:

$$1/c^{w,ss} = (fV)^{w,ss} \tag{2.68}$$

Since we must now distinguish between pure and mutually saturated solvents, the superscripts and subscripts used in the development of the new equivalent of Eq. (2.49), become rather complicated. Eq. (2.67) is assumed to hold true for solute-saturated solvents, and so, from Eqs. (2.67) and (2.68), we have

$$\log K_{ow} = \log(f^{*w,ss}/f^{*oct,ss}) = \log(f^{*w,ss}/f^{*oct,ss} c^{w,ss} f^{w,ss} V^{w,ss}) \tag{2.69}$$

All the quantities in Eq. (2.69) refer to the *solute*. It will be easier to discuss Eq. (2.69) when it is recast into the mole fraction basis. The operative relation is Eq. (2.48). Using Eq. (2.48), we can write

$$f^{*w,ss} = \gamma^{*w,ss} V_{*w}/V \tag{2.70}$$

Table 2.9 Enthalpies of transfer (kJ/mol) from neat to water-saturated octanol

Compound	°C	$\Delta_{tr}H(oct \to {}^*oct)$	Method[†]	Reference
Methanol	25	−1.2	IC	Riebesehl and Tomlinson (1986)
Ethanol	25	0.82	IC	Riebesehl and Tomlinson (1986)
1-Propanol	25	0.65	IC	Riebesehl and Tomlinson (1986)
1-Butanol	25	0.52	IC	Riebesehl and Tomlinson (1986)
1-Pentanol	25	0.55	IC	Riebesehl and Tomlinson (1986)
1-Hexanol	25	0.32	IC	Riebesehl and Tomlinson (1986)
Hexane	25	0.24	IC	Bernazzani et al. (1995)
Heptane	25	0.15	IC	Bernazzani et al. (1995)
Octane	25	0.31	IC	Bernazzani et al. (1995)
Acetone	25	−3.0	IC	Bernazzani et al. (1995)
2-Butanone	25	−1.7	IC	Bernazzani et al. (1995)
2-Hexanone	25	−1.5	IC	Bernazzani et al. (1995)
m-Alkoxyphenols				
Methyl	30	0.72	BC	Beezer et al. (1983)
Propyl	30	1.1	BC	Beezer et al. (1983)

{IC = isoperibol bath calorimetry; BC = batch calorimetry.

$$f^{*oct,ss} = \gamma^{*oct,ss} V_{*oct}/V \tag{2.71}$$

$$f^{w,ss} = \gamma^{w,ss} V_{w,ss}/V \tag{2.72}$$

Eqs. (2.70)–(2.72) can now be substituted into Eq. (2.69). If the simplifying assumptions are made that

$$V_{*w} = V_{w,ss} \tag{2.73}$$

(i.e., the density of water is unaffected by dissolved octanol), then Eq. (2.69) becomes

$$\log K_{ow} = -\log c^{w,ss} - \log(V_{*oct}\gamma^{*oct,ss}) + \log(\gamma^{*w,ss}/\gamma^{w,ss}) \tag{2.74}$$

Eq. (2.74) is to be compared to Eq. (2.49). They differ essentially by the factor $\log(\gamma^{*w,ss}/\gamma^{w,ss})$. In other words, if the (Henrian) solute activity coefficient at saturation in pure water is the same as that in the octanol-saturated water phase in a shake-flask experiment, then the two equations are functionally equivalent.

An approximate indication of the magnitude of $\log(\gamma^{*w}/\gamma^{w,ss})$ may be ascertained from Table 2.6. From Eqs. (2.44) and (2.46), we infer

$$\log (\gamma^{*w,ss}/\gamma^{w,ss}) = \log (c^{w,ss}/c^{*w,ss}) \tag{2.75}$$

Table 2.10 lists values of this ratio for the compounds in Table 2.6. It is evident that, in some cases, the activity coefficient ratio can introduce unexpected deviations into a simple log K_{ow}–log solubility plot.

Eq. (2.74) indicates that, neglecting the activity coefficient ratio, the intercept of the simple correlation plot should be $-\log(\gamma^{*oct} V_{*oct})$. In this term V_{*oct} is independent of solute. According to the data analysis of Mackay *et al.* (1980), γ^{*oct} is only weakly dependent on solute structure (compared to, say, Henrian activity coefficients in water). The possible exceptions to this rule are: very hydrophobic compounds and strongly hydrogen-bonded species, such as acids (Mackay *et al.*, 1980). Henrian activity coefficients of various compounds were measured by GLC and head-space gas chromatography in anhydrous octanol (Dallas and Carr, 1992; Tse and Sandler, 1994; Wasik *et al.*, 1981) and in water-saturated octanol (Dallas and Carr, 1992).

The effect of the water in wet octanol upon solute activity coefficients has not received much attention. Dallas and Carr (1992) measured activity coefficients of 20 solutes in wet and dry octanol by the head-space method. Kristl and

Table 2.10 Activity coefficient ratios of various solutes in pure and octanol-saturated water at 25°C (from Table 2.6)

Compound	$\log(\gamma^{*w,ss}/\gamma^{w,ss})$
Naphthalene	0.04
Biphenyl	0.04
1-Methylfluorene	0.02
Anthracene	0.17
Chrysene	−0.74
2,3-Benzofluorene	0.03
4-PCB	0.02
4,4'-PCB	0.26
m-Methoxyphenol	0.11
m-Propoxyphenol	0.14
Acyclovir	0.01
Deoxyacyclovir	0.00
Diacetyldeoxyacyclovir	0.02

Vesnaver (1995) reported activity coefficients of acyclovir derivatives in wet and dry octanol from solubility measurements. The data do not permit general conclusions to be drawn.

6.5 RELATIONSHIPS WITH OTHER SOLVENT-PAIRS

Although the octanol–water pair is the most extensively used for expressing biologically relevant partition coefficients, it is in fact a relative newcomer. A number of solvents have been used or considered previously (Rekker and Mannhold, 1992; Smith et al., 1975): hydrocarbons, halogenated hydrocarbons, alkanols, ethers, ketones and esters, principally of small carbon number. Recently, propyleneglycol dinonanoate/water was proposed as a partitioning standard (Leahy et al., 1989).

With the accumulation of enough partition coefficient data for different solvent/water pairs, the question arises, whether or to what extent they are related. Collander suggested (1951) that there might be a linear relationship (for chemically similar solvents) such as

$$\log K_{ow} = a \log K_{sw} + b \tag{2.76}$$

where sw represents another solvent–water system and a and b are empirical constants. A preliminary investigation was made to determine the applicability of Eq. (2.76) with a large number of solvents and solutes (Leo et al., 1971), distinguishing primarily between hydrogen-donor and -acceptor solutes. It was conjectured (Leo et al., 1971) that, in the absence of solute–solvent hydrogen bonding, the slope of Eq. (2.76) would be unity, and that the intercept would reflect solute–solvent interaction. This suggestion was taken up by Seiler (1974) who derived the relation

$$\log K_{ow} - \log K_{cw} = I_H + b \tag{2.77}$$

where cw refers to the cyclohexane-water system and I_H represents a hydrogen-bonding contribution (values for 21 functional groups were established). In Seiler's derivation, $\log S_{cw}$ is a composite function normalizing alkane–water partition coefficients to the cyclohexane–water system. The LHS of Eq. (2.77) has been called the '$\Delta \log P$ parameter' (Abraham et al., 1994; Leahy et al., 1989) — here renamed '$\Delta \log K$ parameter' — and elucidation of its significance has been the object of some study.

Since partition coefficients are directly related to activity coefficients (Eq. 2.42), the $\Delta \log K$ parameter can be written

$$\log K_{ow} - \log K_{cw} = \log(f^c/f^{oct}) \tag{2.78}$$

$$= \log(\gamma^c/\gamma^{oct}) + \log(V_c/V_{oct}) \tag{2.79}$$

using Eq. (2.48). This is the approach taken by Hawker (1994), who expressed the activity coefficients in Eq. (2.79) in terms of the Hildebrand solubility parameter (δ). Now δ is a useful parameter for describing the hydrophilic-lipophilic balance for certain purposes (Schott, 1995), but since its use is restricted to regular solutions, its applicability is circumscribed (Hawker, 1994).

At this point it may be recognized that the $\Delta \log K$ parameter is actually a straightforward Gibbs energy of transfer. Thus, applying Eq. (2.61) to both solvent pairs, we have

$$\log K_{ow} - \log K_{cw} = \Delta_{tr}G(\text{oct} \to \text{c})/2.303RT + \log(V_c/V_{oct}) \qquad (2.80)$$

and the common solvent, water, drops out of the relation.

It would appear (Abraham, 1984) that the $\Delta \log K$ parameter represents a simple linear function for homologous series of compounds. For the octanol–water (ow) and hexadecane–water (hw) pairs, Schantz and Martire (1987), using a lattice model theory, derived the relation

$$\log K_{ow} - \log K_{hw} = an_{CH_2} + b \qquad (2.81)$$

where n_{CH_2} is the number of methylene groups in the chain.

A recent predictive scheme for partition coefficients has been quite successful: linear solvation energy relationships (LSER, El Tayar *et al.*, 1991; Abraham *et al.*, 1994; Leahy *et al.*, 1992). A more detailed account of LSER will be given in Chapter 5; here the treatment of Abraham *et al.* (1994) will be used for illustration. The solvent–water partition coefficient $\log K_{sw}$ is a linear function of a number of solute properties:

$$\log K_{sw} = c + rR_2 + s\pi_2^H + 2\alpha_2^H + b\beta_2^H + vV_x \qquad (2.82)$$

where c, r, s, a, b and v are fitting parameters dependent on only the solvent, and the remaining variables are the solute properties (indicated by subscript 2):

R_2: excess molar refraction

π_2^H: dipolarity/polarizability

α_2^H: effective hydrogen-bond basicity

β_2^H: effective hydrogen-bond acidity

V_x: a characteristic molar volume

For the 288 compounds for which both $\log K_{ow}$ and $\log K_{hw}$ were available (Abraham *et al.*, 1994), the $\Delta \log K$ parameter of Eq. (2.81) was

$$\log K_{ow} - \log K_{hw} = -0.72 - 0.093R_2 + 0.528\pi_2^H + 3.65\alpha_2^H + 1.396\beta_2^H - 0.521V_x$$
$$(2.83)$$

It is not immediately apparent from Eqs (2.81) and (2.83) what the $\Delta \log K$ parameter is. A few such plots are given in Figure 2.4 for some homologous

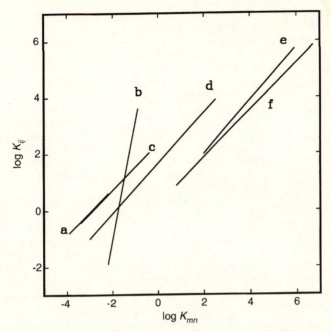

Figure 2.4 The $\Delta \log K$ parameter for homologous series in different solvent pairs. Solvent pairs are identified in Table 2.11: a, 3-pyridine alkylamides (Repond *et al.*, 1987); b, 1-alkanols (Schantz *et al.*, 1988); c, alkanoic acids (Abraham *et al.*, 1994); d, 1-alkanols (Abraham *et al.*, 1994); (e), alkylbenzenes (Abraham *et al.*, 1994); f, n-alkanes (Abraham *et al.*, 1994).

series (Abraham *et al.*, 1994; Repond *et al.*, 1987; Schantz *et al.*, 1988). A more complete comparison is presented in Table 2.11. Within experimental uncertainty, the relations are all linear; the reason can readily be seen from Eq. (2.83) and homologous series properties (Abraham *et al.*, 1994). Thus in homologous series, either the solute parameters are constant or vary linearly with carbon number. Evidently the relationships for the homologous series *n*-alkanes, 1-alkenes, 1-alkynes, 1-bromoalkanes, are all very similar, but the introduction of polar groups brings in other factors.

6.6 TREATMENT OF IONIZABLE COMPOUNDS

The definition of the partition coefficient, Eq. (1.1), is strictly applicable only for those cases in which the same species is considered in both solvents. It is commonplace, however, to find K_{ow} data for compounds for which this proviso may not be true. Three instances of this apparent anomaly will be considered here: (1) weak organic acids (or bases), (2) quaternary ammonium (phosphonium, sulphonium) compounds and (3) zwitterions.

Table 2.11 The $\Delta \log K$ parameter for some homologous series, for some solvent-water pairs at 25°C, $\log K_{iw} = a \log K_{jw} + b$

i*	j	a	b	Homologous series (number of compounds)	Reference
oct	hex	0.84	0.21	alkanes (9)	Abraham *et al.* (1994)
oct	hex	0.94	0.08	alkanes (4)	Schantz and Martire (1987)
oct	hex	0.83	0.33	1-bromoalkanes (8)	Abraham *et al.* (1994)
oct	hex	0.98	−0.06	1-bromoalkanes (4)	Schantz and Martire (1987)
oct	hex	0.89	1.69	1-alkanols (9)	Abraham *et al.* (1994)
oct	hex	0.95	1.58	1-alkanols (4)	Schantz and Martire (1987)
oct	hex	0.95	0.13	alkylbenzenes (7)	Abraham *et al.* (1994)
oct	hex	0.94	0.08	alkylbenzenes (5)	Schantz and Martire (1987)
oct	hex	0.82	2.37	alkanoic acids (5)	Abraham *et al.* (1994)
oct	hex	0.92	−0.01	1-alkenes (8)	Abraham *et al.* (1994)
oct	hex	1.05	−0.51	1-alkenes (4)	Schantz and Martire (1987)
oct	hex	0.89	0.28	1-alkynes (5)	Abraham *et al.* (1994)
hex	m	4.22	7.39	1-alkanols (6)	Schantz *et al.* (1987)
oct	b	0.82	2.39	3-pyridinealkanamides (4)	Repond *et al.* (1987)

*oct = octanol; hex = hexadecane; m = methanol; b = dibutylether

Weak acids or bases. Published analyses of the partitioning of ionized or ionizable compounds between immiscible solvents sometimes contain ambiguities. A clarifying discussion is given by Taylor (1990). The following treatment, though not absolutely rigorous, will suffice to set out the principles involved (van der Giesen and Janssen, 1982; Shim *et al.*, 1981; Terada *et al.*, 1981).

For an organic acid HA (a parallel analysis holds for bases, *pari passu*) in aqueous solution, the thermodynamic dissociation constant is

$$K_a = a(H^+)a(A^-)/a(HA) \approx [H^+][A^-]/[HA] \qquad (2.84)$$

where the as are activities and HA is the undissociated acid. (The thermodynamic dissociation constant is unitless.) The partition coefficient is

$$K_{ow} = [HA]^{oct}/[HA]^w \qquad (2.85)$$

It is generally assumed (this is not absolutely true) that separate ions do not enter the organic phase in solvent/water partitioning. Hence, in order to obtain the true partition coefficient of HA, one most assure that ionization of the acid is suppressed in the aqueous layer (e.g. by using 0.1M HCl). The situation is illustrated in Figure 2.5(a). If, in this case, HA is allowed to dissociate at pH not too different from pK_a, the partition coefficient *as measured* will be

$$K_{app} = [HA]^{oct}/([HA]^w + [A^-]^w) \qquad (2.86)$$

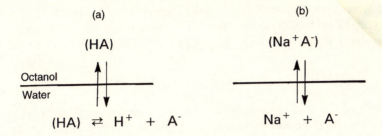

Figure 2.5 Partitioning of a weak acid: (a) pH near pK_a, (b) pH $\gg pK_a$.

where K_{app} is an *apparent* partition coefficient (also called distribution coefficient, D). Under these circumstances, the K_{app} may be corrected for (partial) ionization as follows. From Eqs (2.84)–(2.86),

$$K_{ow}/K_{app} = 1 + [A^-]^w/[HA]^w \tag{2.87}$$

$$[A^-]^w/[HA]^w = K_a/[H^+]^w = \text{antilog}(pH - pK_a) \tag{2.88}$$

since, by definition, $[H^+] = 10^{-pH}$ and $\log K_a = -pK_a$. Thus, from Eqs (2.87) and (2.88),

$$\log K_{ow} = \log K_{app} + \log[1 + \text{antilog}(pH - pK_a)] \tag{2.89}$$

Eq. (2.89) is often used in the partition coefficient literature; it use is legitimate, however, only at a pH not too far from pK_a.

Eq. (2.89) is the basis of a graphical method for determining K_{ow} from K_{app} measured at different pH values. Eq. (2.89) may be rearranged to give

$$1/K_{app} = [1 + \text{antilog}(pH - pK_a)]/K_{ow} \tag{2.90}$$

$$= 1/K_{ow} + \text{antilog}(pH - pK_a)/K_{ow} \tag{2.91}$$

$$= 1/K_{ow} + K_a/K_{ow}[H^+]^w \tag{2.92}$$

or

$$[H^+]^w/K_{app} = [H^+]^w/K_{ow} + K_a/K_{ow} \tag{2.93}$$

According to Eq. (2.93), a plot of the LHS vs $[H^+]^w$ should yield a straight line, of slope $1/K_{ow}$ (Ezumi and Kubota, 1980; Johansson and Gustavii, 1976).

When pH $\gg pK_a$, the acid exists as its salt and the situation is that of Figure 2.5(b) (the counter-ion is here assumed to be Na^+). The apparent partition coefficient is

$$K_{app} = [(Na^+A^-)]^{oct}/[A^-]^w \tag{2.94}$$

where (Na^+A^-) is an ion-pair (see following).

Quaternary ammonium salts. In this category are non-hydrolyzable salts, in which one or both of the cation and anion may be organic. If the salt is Q^+A^- (Q^+ is the quaternary cation), the ion-pair partition coefficient may be given as

$$K_{ow} = [(Q^+A^-)]^{oct}/[Q^+A^-)]^w \qquad (2.95)$$

i.e., referring to ion-pairs in both solvents. Commonly, what is reported is simply the total concentration in the octanol phase divided by the total concentration in the water phase:

$$K_{app} = [(Q^+A^-)]^{oct}/\{[(Q^+A^-)] + [Q^+]\}^w \qquad (2.96)$$

where $[Q^+]^w = [A^-]^w$ in the absence of any added common-ion salt. The concentrations of ion-pairs and separate ions in the aqueous phase (denominator of Eq. 2.96) are related by the ion-pair association constant K_{ass}

$$K_{ass} \approx [(Q^+A^-)]^w/[Q^+]^w[A^-]^w \qquad (2.97)$$

If K_{ass} is sufficiently small, then Eq. (2.96) assumes the form of Eq. (2.94). Alternatively, if it is sufficiently large, Eq. (2.96) assumes the form of Eq. (2.95).

The ion-pair extraction constant, sometimes represented by E, is

$$E = [(Q^+A^-)]^{oct}/[Q^+]^w[A^-]^w \qquad (2.98)$$

(Shim *et al.*, 1981; Terada *et al.*, 1981). In Eq. (2.98), allowance must be made for the existence of ion-pairs in the aqueous phase.

6.7 ZWITTERIONIC COMPOUNDS

There are several types of organic molecules which have two or more ionized or ionizable groups on the same carbon backbone (Laughlin, 1991). A few examples are given in Table 2.12.

Of particular interest here are amino acids and peptides. Amino acids, e.g. glycine (H_2NCH_2COOH) contain both an amino group and a side chain at the α-position to an acid group. This side chain may itself be ionizable or not. Peptides are end-to-end condensation products of two or more amino acids, e.g., glycyl alanine $H_2NCH_2CONHCH(CH_3)COOH$).

An amino acid such as glycine may exist in aqueous medium in several forms, viz., $H_3N^+CH_2COOH$, $N_2NCH_2COO^-$ or $H_3N^+CH_2COO^-$. These forms are most in evidence in acidic, basic or neutral media, respectively. The apparent partition coefficients of amino acids and peptides would therefore be expected to be pH-dependent. This is, in fact, what is found experimentally, for amino acids with un-ionizable (El Tayar *et al.*, 1992; Fauchère and Pliška,

Table 2.12 Examples of zwitterionic compounds

Name or type	Structure
Glycine	H_2NCH_2COOH
A bolaform electrolyte	$Br^-(CH_3)_3N^+(CH_2)_4N^+(CH_3)_3Br^-$
Betaine chloride	$Cl^-(CH_3)_3N^+CH_2COOH$
p-Aminobenzoic acid	$H_2NC_6H_4COOH$
A phosphoniosulphate compound	$Cl^-(CH_3)_3P^+CH(CH_3)CH_2SO_4^-Na^+$

1983) or ionizable side chains (Fauchère and Pliška, 1983; Yunger and Cramer, 1981) and for peptides (Akamatsu et al., 1989). When end groups of peptides are protected by acylation or amidization, the pH sensitivity corresponds to that of the side chain (Kim and Szoka, 1991). As a rule, the K_{app} vs. pH curve passes through a maximum, more or less flat, around the isoelectric point. This behaviour is consistent with the fact that an amino acid such as glycine in aqueous medium has a *net* charge of zero at the *isoelectric point* (usually taken as the mean of the pK_a values of the acid and amino groups). The isoelectric points of most amino acids or peptides lie between 5.2 and 5.7 (Akamatsu and Fujita, 1991; Jencks and Regenstein, 1976). Since the pH profile of K_{app} is usually quite flat between pH 5 and 7, these compounds are considered to be 'neutral' in this interval, and hence partition into octanol in the same form. The K_{ow} of a large number of amino acids, peptides and similar compounds has been measured and reported, at or near pH 7.

7 LOG K_{ow} AND OTHER PHYSICOCHEMICAL PROPERTIES

Log K_{ow} is a Gibbs energy transfer function and is an additive-constitutive function of molecular structure. It is not surprising, therefore, to find that it may be correlated more or less closely with other such physicochemical properties. Here two correlation studies, of rather broad scope, are described; more restricted log K_{ow} predictive methods are examined in Chapter 5.

Factor (principal component) analysis was used (Cramer, 1980) to deduce a linear correlation of log K_{ow} with other physicochemical properties (water solubility, molar refractivity, melting point, boiling point, molecular weight and 15 others). This BC(DEF) parameter analysis reproduced log K_{ow} data of 114 liquid compounds with a standard deviation of 0.08. Such an approach to correlation will, of course, be successful as the number of parameters increases, but the exercise yields little or no insight into fundamental relationships.

Straightforward regression was performed on the log K_{ow} data of 301 compounds (solids, liquids and gases) with each of nine physicochemical properties (Mailhot and Peters, 1988). In addition to six properties considered by Cramer (1980), three others were included: parachor, connectivity index and surface area. The most successful predictor was water solubility. This result is to be expected, since there is a thermodynamic relationship between $\log K_{ow}$ and solubility.

8 PARTITION COEFFICIENTS OF ISOMERS AND TAUTOMERS

Isomers are different compounds with the same molecular formula. Two principal kinds of isomers are usually distinguished: constitutional (structural) isomers and stereoisomers.

Constitutional isomers differ because their atoms are connected in a different order. Table 2.13 gives $\log K_{ow}$ data of constitutional isomers of $C_4H_{10}O$. Constitutional isomers are distinct chemical species having different melting points, boiling points, densities, refractive indices, etc. Their partition coefficients also are different, as Table 2.13 shows.

In stereoisomers, the atoms are connected in the same order, but the molecules differ in arrangement of their atoms in space. Two types of stereoisomers are usually distinguished. *Enantiomers* are stereoisomers that are mirror images of each other. (Enantiomers display optical activity in solution and contain one or more stereogenic (asymmetric) carbon atoms.) For example, (R)- and (S)- 2-butanol are enantiomers, as are D- and L-forms of amino acids. Enantiomers have identical melting points, boiling points, refractive indices, etc. Careful systematic partition coefficients measurements have not been reported for enantiomers, but it may be concluded that partition coefficients of enantiomers are identical also.

Diastereomers are stereoisomers which are *not* mirror images of each other. Simple examples are: the cis- and trans-isomers of 2-butene,

Table 2.13 Partition coefficients of constitutional isomers of $C_4H_{10}O$ (Sangster, 1993)

Compound	$\log K_{ow}$
Methylpropyl ether	1.21
Methylisopropyl ether	1.01*
Diethyl ether	0.89
1-Butanol	0.84
2-Butanol	0.65
Isobutanol	0.76
tert-Butanol	0.35

*Calculated value

1,2-dichloroethylene and 1,2-dimethylcyclopentane. Diastereomers are distinct chemical species and have different melting and boiling points, refractive indices, etc., although the values are usually close to each other. Measured $\log K_{ow}$ data for (E)- and (Z)- 2-butene and (E)- and (Z)- 1,2-dichloroethylene are (2.31, 2.33) and (2.04, 1.86) respectively (Sangster, 1993).

A careful partition coefficient study of diastereomers was reported by Kurihara *et al.* (1980, 1983) on polychlorinated cyclohexenes and cyclohexanes. These compounds have insecticidal activity. The best known of these are the hexachlorocyclohexanes (Table 2.14), of which lindane (the γ-isomer) is most familiar.

According to Eq. (2.42), K_{ow} of organic compounds is equivalent to the ratio of Henrian activity coefficients of the solute in the two solvents. For similar compounds, the activity coefficient in octanol is rather insensitive to structure (Dallas and Carr, 1992; Mackay *et al.*, 1980; Schantz and Martire, 1987; Wasik *et al.*, 1981). Most of the dependence of K_{ow} on structure for isomers, then, would be attributed to different interactions with water. Evidently enantiomers in solution are identically solvated and partition coefficients are consequently identical.

Tautomers may be considered to be two forms of the same molecule in mobile equilibrium. The equilibrium may depend upon the existence and type of solvent. A well-known example is acetylacetone (keto-enol tautomerism):

$$CH_3COCH_2COCH_3 \longleftrightarrow CH_3COCH=C(OH)CH_3 \qquad (2.99)$$

In the pure state, most of this substance exists as the enol, but in the octanol–water system this equilibrium is probably different (Hansch and Leo, 1995). A similar compound, ethyl acetoacetate ($CH_3COCH_2COOCH_2CH_3$) is almost all in the keto form; the same is true for heterocyclic aromatic compounds such as 2-pyridinone, 2-pyridinethione and barbituric acid. Tautomerism, of course, is not usually of crucial importance in partition coefficient measurements. It requires attention, however, in the elaboration of schemes for calculating $\log K_{ow}$ from structures (Chapter 5).

Table 2.14 Partition coefficients of 1, 2, 3, 4, 5, 6-hexachlorocyclohexane isomers (Kurihara and Fujita, 1983)

Isomer	Configuration*	$\log K_{ow}$
α	aaeeee	3.82
β	eeeee	3.80
γ	aaaeee	3.72
γ	aeeeee	4.14

*a=axial, e=equatorial. The designation *configuration* is used here rather than *conformation* (Roberts and Caserio, 1967)

REFERENCES

Abraham, M. H. (1984) *J. Chem. Soc. Faraday Trans.* **80**, 153–81.

Abraham, M. H. (1988) *J. Chem. Soc. Faraday Trans.* **84**, 1985–2000.

Abraham, M. H., Chadha, H. S., Whiting, G. S. and Mitchell, R. C. (1994) *J. Pharm. Sci.* **83**, 1085–1100.

Akamatsu, M. and Fujita, T. (1992) *J. Pharm. Sci*, **81**, 164–74.

Akamatsu, M., Yoshida, Y., Nakamura, H., Asao, M., Iwamura, H. and Fujita, T. (1989) *Quant. Struct.-Act. Relat.* **8**, 195–203.

Anliker, R. and Moser, P. (1987) *Ecotoxicol. Environ. Saf.* **13**, 43–52.

Apelblat, A. (1983) *Ber. Bunsenges. Phys. Chem*, **87**, 2–5.

Apelblat, A. (1990) *Ber. Bunsenges. Phys. Chem*, **94**, 1128–34.

Beezer, A. E., Hunter, W. H. and Storey, D. E. (1980) *J. Pharm. Pharmacol*, **32**, 815–19.

Beezer, A. E., Hunter, A. E. and Storey, D. E. (1983) *J. Pharm. Pharmacol.* **35**, 350–7.

Beezer, A. E., Lima, M. C. P., Fox, G. G., Arriaga, P., Hunter, W. H. and Smith, B. V. (1987) *J. Chem. Soc. Faraday Trans. I* **83**, 2705–7.

Bernazzani, L., Cabani, S., Conti, G. and Mollica, V. (1995) *J. Chem. Soc. Faraday Trans.* **91**, 649–55.

Berti, P., Cabani, S., Conti, G. and Mollica, V. (1986) *J. Chem. Soc. Faraday Trans. I* **82**, 2547–56.

Berti, P., Cabani, S., Conti, G. and Mollica, V. (1987) *Thermochimica Acta* **122**, 1–8.

Betageri, G. V., Nayernama, A. and Habib, M. J. (1996) *Int. J. Pharm. Adv.* **1**, 310–19.

Boustead, I. and Hancock, G. F. (1979) *Handbook of Industrial Energy Analysis*, Ellis Horwood, Chichester.

Bowman, B. T. and Sans, W. W. (983) *J. Environ. Sci. Health B* **18**, 667–83.

Brodin, A., Sandin, B. and Faijerson, B (1976) *Acta Pharm. Suec.* **13**, 331–52.

Butler, J. A. V., Thomson, D. W. and McLennan, W. N. (1993) *j. Chem. Soc.* 674–86.

Cabani, S., Conti, G., Mollica, V. and Berti, P. (1990) *Thermochimica Acta* **162**, 195–201.

Cabani, S., Conti, G., Mollica, V. and Bernazzani, L. (1991) *J. Chem. Soc. Faraday Trans.* **87**, 2433–42.

Chase, M. W., Davies, C. A., Downey, J. R., Frurip, D. J., McDonald, R. A. and Syverud, A. N. (1985) *J. Phys. Chem. Ref. Data* **14**(Suppl. No. 1), 1–1856.

Chiogioji, M. H. (1979) *Industrial Energy Conservation*, Marcel Dekker, New York.

Chiou, C. T., Freed, V. H., Schmedding, D. W. and Kohnert, R. L. (1977) *Environ. Sci. Technol.* **11**, 475–8.

Chiou, C. T., Schmedding, D. W. and Manes, M. (1982) *Environ. Sci. Technol.* **16**, 4–10.

Collander, R. (1951) *Acta Chem. Scand.* **5**, 774–80.

Cramer, R. D. (1980) *J. Am. Chem. Soc.* 102, 1837–59.

Dagorn, M., Huet, J. and Burgot, J. L. (1985) *Ann. Pharm. Fr.* **43**, 165–71.

Dallas, A. J. and Carr, P. W. (1992) *J. Chem. Soc. Perkin Trans.* 2, 2155–61.

Dallos, A. and Liszi, J. (1995) *J. Chem. Thermodyn.* **27**, 447–8.

DeVoe, H., Miller, M. M. and Wasik, S. P. (1981) *J. Res. Nat. Bur. Stand. (U.S.)* **86**,361–6.

El Tayar, N., Tsai, R.-S., Testa, B., Carrupt, P.-A. and Leo, A. (1991) *J. Pharm. Sci.* **80**, 590–8.

El Tayar, N., Tsai, R.-S., Carrupt, P.-A. and Testa, B. (1992) *J. Chem. Soc. Perkin Trans*, 2, 79–84.

Ezumi, K. and Kubota, t. (1980) *Chem. Pharm. Bull.* **28**, 85–91.

Fauchère, J.-L. and Pliška, V. (1983) *Eur. J. Med. Chem.-Chim. Ther.* **18**, 369–75.

van der Giesen, W. F. and Janssen, L. H. M. (1982) *Int. J. Pharm.* **12**, 231–49.

Gmehling, J. and Onken, U. (1977) *Vapor–liquid Equilibrium Data Collection*, Vol. 1, Part 1, Deutsche Gesellschaft für Chemisches Apparatewesen, Frankfurt/Main.

Gmehling, J., Onken, U. and Arlt, W. (1981) *Vapor–liquid Equilibrium Data Collection*, Vol. 1, Part 1a, Deutsche Gesellschaft für Chemisches Apparatewesen, Frankfurt/Main.

Guggenheim, E. A. (1986) *Thermodynamics: an advanced treatment for chemists and physicists*, 8th edition, North-Holland, Amsterdam.

Guy, R. H. and Honda, D. H. (1984). *Int. J. Pharm.* **19**, 1239–37.

Hallén, D., Nilsson, S.-O., Rothschild, W. and Wadsö, I. (1986) *J. Chem. Thermodyn.* **18**, 429–42.

Hansch, C. and Leo, A. (1995) *Exploring QSAR: fundamentals and applications in chemistry and biology*, Chap. 5, American Chemical Society, Washington.

Hansch, C., Quinlan, J. E. and Lawrence, G. L. (1968) *J. Org. Chem.* **33**, 347–50.

Hatanaka, T., Inuma, M., Sugitayashi, K. and Morimoto, Y. (1990) *Chem. Pharm. Bull.* **38**, 3452–9.

Hawker, D. (1994) *Toxicol. Environ. Chem.* **45**, 87–95.

Hill, D. J. T. and White, L. R. (1974) *Aust. J. Chem.* **27**, 1905–16.

Isnard, P. and Lambert, S. (1989) *Chemosphere* **18**, 1837–53.

Jencks, W. P. and Regenstein, J. (1976) In Vol. 1 of CRC *Handbook of Biochemistry and Molecular Biology*, 3rd edition, ed. G. D. Fasman, CRC Press, Boca Raton, pp. 305–51.

Johansson, P.-A. and Gustavii, K. (1976) *Acta Pharm. Suec.* **13**, 407–29.

Kim, A. and Szoka, F. C. (1992) *Pharm. Res.* **9**, 504–14.

Kinkel, J. F. M., Tomlinson, E. and Smit, P. (1981) *Int. J. Pharm.* **9**, 121–36.

Klotz, I. M. and Rosenberg, R. M. (1986) *Chemical Thermodynamics: basic theories and methods*, 4th edition, Addison-Wesley, Reading, MA.

Korenman, Ya. I. and Udalova, V. Yu. (1974) *Russ. J. Phys. Chem. (Engl. Transl.)* **48**, 708–11.

Kosanovic, D., Dumanovic, D. and Jovanovic, J. (1988) *J. Serb. Chem. Soc,* **53**, 559–63.

Kramer, C.-R. and Henze, U. (1990) *Z. Phys. Chem. (Leipzig)* **271**, 503–13.

Kristl, A. (1996) *J. Chem. Soc. Faraday Trans.* **92**, 1721–4.

Krisl, A. and Vesnaver, G. (1995) *J. Chem. Soc. Faraday Trans.* **91**, 995–89.

Kurihara, N. and Fujita, T. (1983) *Bull. Inst. Chem. Res., Kyoto Univ.* **61**, 89–95.

Kurihara, N., Yamakawa, K., Fujita, T. and Nakajima, M. (1980) *J. Pestic. Sci.* **5**, 93–100.

Laughlin, R. G. (1991) *Langmuir* **7**, 842–7.

Leahy, D. E. (1986) *J. Pharm. Sci.* **75**, 629–36.

Leahy, D. E., Taylor, P. J. and Wait, A. R. (1989) *Quant. Struct.-Act. Relat.* **8**, 17–31.

Leahy, D. E., Morris, J. J., Taylor, P. J. and Wait, A. R. (1992) *J. Chem. Soc. Perkin Trans.* 2, 723–31.

Leo, A., Hansch, C. and Elkins, D. (1971) *Chem. Rev.* **71**, 525–616.

Lewis, G. N. and Randall, M. (1961) *Thermodynamics*, ed. K. S. Pitzer and L. Brewer, McGraw-Hill, New York.

Lyzlova, R. V., Zaiko, L. N. and Susarev. M. P. (1979) *Zh. Prikl. Khim. (Leningrad)* **52**, 551–5.

Mackay, D., Bobra, A. and Shiu, W.-Y. (1980) *Chemosphere* **9**, 701–11.

Mailhot, H. and Peters, R. H. (1988) *Environ. Sci. Technol.* **22**, 1479–88.

Miller, M. M., Ghodbane, S., Wasik, S. P. Tewari, Y. B. and Martir, D. E. (1984) *J. Chem. Eng. Data* **29**, 184–90.

Miller, M. M., Wasik, S. P., Huang, G.-L., Shiu, W.-Y. and Mackay, D. (1985) *Environ. Sci, Technol.* **19**, 522–9.

Niimi, A. J. (1991) *Water Res.* **25**, 1515–21.

Nilsson, S.-O. (1986a) *J. Chem. Thermodyn.* **18**, 877–84.

Nilsson, S.-O. (1986b) *J. Chem. Thermodyn.* **18**, 1115–23.

Opperhuizen, A., Serné, P. and van der Steen, J. M. D. (1988) *Environ. Sci. Technol.* **22**, 286–92.

Patil, G. S. (1991) *Chemosphere* **22**, 723–38.

Patil, G. S. and Bora, M. (1994) *Toxicol. Environ. Chem.* **45**, 205–9.

Pfeffer, T., Löwen, B. and Schulz, S. (1995) *Fluid Phase Equil.* **106**, 139–67.

Pinsuwan, S. Li, A. and Yalkowsky, S. H. (1995) *J. Chem. Eng. Data* **40**, 623–6.

Quigley, J. M., Fahelelbom, K. M. S., Timoney, R. F. and Corrigan, O. I. (1990) *Int. J. Pharm.* **58**, 107–13.

Ragone, D. V. (1995) *Thermodynamics of Materials*, 2nd edition, John Wiley and Sons, New York.

Reid, C. E. (1990) *Chemical Thermodynamics*, McGraw-Hill, New York.

Reid, R. C., Prausnitz, J. M. and Poling, B. E. (1987) *The Properties of Gases and Liquids*, 4th edition, McGraw-Hill, New York.

Rekker, R. F. and Mannhold, R. (1992) *Calculation of Drug Lipophilicities*, VCH Verlagsgesellschaft, Weinheim.

Repond, C., Mayer, J. M., van de Waterbeemd, H., Testa, B. and Linert, W. (1987) *Int. J. Pharm.* **38**, 47–57.

Riebesehl, W. and Tomlinson, E. (1983) *J. Chem. Soc. Faraday Trans. I* **79**, 1311–16.

Riebesehl, W. and Tomlinson, E. (1984) *J. Phys. Chem*, **88**, 4770–5.

Riebesehl, W. and Tomlinson, E. (1986) *J. Solution Chem.* **15**, 141–50.

Riebesehl, W., Tomlinson, E. and Grünbauer, H. J. M. (1984) *J. Phys. Chem.* **88**, 4775–9.

Riebesehl, W., Tomlinson, E. and Niemel, P. R. (1985) *J. Solution Chem.* **14**, 699–707.

Roberts, J. D. and Caserio, M. C. (1967) *Modern Organic Chemistry*, W. A. Benjamin, New York.

Rogers, J. A. and Wong, A. (1980) *Int. J. Pharm.* **6**, 339–48.

Rowlinson, J. S. and Swinton, F. L. (1982) *Liquids and Liquid Mixtures*, Butterworths, London.

Sangster, J. (1989) *J. Phys. Chem. Ref. Data* **18**, 1111–229.

Sangster, J. (1993) *LOGKOW—a databank of evaluated octanol-water partition coefficients*, Sangster Research Laboratories, Montreal.

Schantz, M. M. and Martire, D. E. (1987) *J. Chromatogr.* **391**, 35–51.

Schantz, M. M., Barman, B. N. and Martire, D. E. (1988) *J. Res. Nat. Bur. Stand. (U.S.)* **93**, 161–73.

Schott, H. (1995) *J. Pharm. Sci*, **84**, 1215–22.

Seiler, P. (1974) *Eur. J. Med. Chem.-Chim. Ther*, **9**, 473–9.

Shim, C. K., Nishigaki, R., Iga, T. and Hanano, M. (1981) *Int. J. Pharm.* **8**, 143–51.

Smith, R. N., Hansch, C. and Ames, M. M. (1975) *J. Pharm. Sci*, **64**, 599–606.

Taylor, P. J. (1990) 'Hydrophobic properties of drugs', in C. Hansch, gen. ed., *Comprehensive Medicinal Chemistry*, Vol. 4, Pergamon Press, New York, pp. 241–94.

Terada, H., Keiko, K., Yoshikawa, Y. and Kametani, F. (1981) *Chem. Pharm. Bull.* **29**, 7–14.

Tewari, Y. B., Miller, M. M., Wasik, S. P. and Martire, D. E. (1982) *J. Chem. Eng. Data* **27**, 451–4.

Tomlinson, E. (1983) *Int. J. Pharm.* **13**, 115–44.

Tomlinson, E., Riebesehl, W. and Grünbauer, H. J. M. (1986) *Pure Appl. Chem.* **58**, 1573–84.

Tse, G. and Sandler, S. I. (1994) *J. Chem. Eng. Data* **39**, 354–7.

Wasik, S. P., Tewari, Y. B., Miller, M. M. and Martire, D. E. (1981) *Octanol-water Partition Coefficients and Aqueous Solubilities of Organic Compounds*, NBSIR 81-2406, U. S. Department of Commerce, Washington.

Yalkowsky, S. H., Valvani, S. C. and Roseman, T. J. (1983) *J. Pharm. Sci.* **72**, 866–70.

Yunger, L. M. and Cramer, R. D. (1981) *Mol. Pharmacol.* **20**, 602–8.

CHAPTER 3

Experimental Methods of Measurement

It has been said (Taylor, 1990) that '. . . despite some notable recent advances, log P measurement remains in the horse-and-buggy era. . .'. It is true that there is as yet no method which has all the advantages of the shake-flask procedure and none of its inconvenience, fuss and bother. In this chapter, a large number of methods will be described; discussion and comparison of these is reserved mostly for Chapter 4. Here, they are described under two subheadings, 'direct' and 'indirect' methods (Sangster, 1989) for convenience. In the present context, *direct* means that one or both of the immiscible phases are analysed quantitatively for solute. *Indirect* means that there is no quantitative analysis. Such a categorization should not be taken to imply any prejudgment on their usefulness or quality.

1 DIRECT METHODS

1.1 SHAKE-FLASK

This is the classical extraction-flask procedure, used in preparative organic chemistry, to isolate a compound (Figure 1.1). For measuring partition coefficients, the solute is dissolved in one phase, and through agitation it becomes distributed between the two phases. After separation, each phase is analysed for solute. Stated in this manner, the method is very simple. In practice, however, it is necessary to pay attention to a number of details, particularly for high accuracy. These details have been discussed with varying degrees of thoroughness (Dearden and Bresnen, 1988; Leo, 1991; Leo *et al.*, 1971; Martin, 1978; Purcell *et al.*, 1973; Rekker, 1977; Taylor, 1990).

Since the concentration of solute used in partition measurements is usually 10^{-3} M or less, the purity of the solvents used becomes important, e.g., for analysis by absorption spectrophotometry. Also, distilled water in equilibrium with atmospheric carbon dioxide becomes distinctly acidic. The solute itself must be sufficiently pure since, if the analytical method cannot distinguish between the solute and an impurity with a different partition coefficient, the result can be noticeably in error (Dearden and Bresnen, 1988). This problem does not arise if both phases are analysed by GLC or HPLC. This was demonstrated (Tewari *et al.*, 1982a) by partitioning a mixture of ethylbenzene and propylbenzene and anlaysing by HPLC. The partition coefficients of both solutes were found to be independent of each other's presence. The concentration of solute must be low enough so that the solutions are effectively Henrian with respect to the solute (organic acids may associate in solution at high enough concentration).

The solute may sometimes be difficult to dissolve initially in water or octanol. A small amount of auxiliary solvent, such as methanol, perhaps with heating, can be used to effect solution. The small amount of auxiliary solvent does not interfere with the partition process.

The solvents are pre-saturated with each other before use; this is particularly important when the exact volumes of phases are required information.

When K_{ow} of an ionizable solute is being measured and the true partition coefficient is desired, the aqueous phase must be buffered or otherwise modified to suppress ionization. For simple organic acids and bases, 0.1 M HCl and 0.1 M NaOH respectively may be used. Otherwise phosphate, citrate, bicarbonate, acetate or tris(hydroxymethyl)aminomethane hydrochloride buffers are available. The buffer capacity, of course, should be adequate for the concentration of solute concerned, and the buffer itself should not be extractable. A phosphate buffer appears to present the least possibility of complicating factors (Wang and Lien, 1980).

Although agitation of the two phases speeds up the attainment of equilibrium, the equilibration step can be accomplished without stirring ('sit flask' experiment, 7–11 days: Schantz *et al.*, 1988). Agitation may be performed by a machine shaker or by end-over-end motion, etc. Shaking times of 2–30 minutes have been recommended. Violent shaking is neither necessary nor advisable, since emulsions may form under these conditions (any surface active compound must, of course, be at concentrations below the critical micelle concentration). After agitation, most of the separation occurs upon standing; this is followed by centrifugation to complete the process. Centrifugation times of up to 2 h at 2000 or 3600 rpm have been recommended.

Separation of the phases for analysis sometimes requires careful manipulation. In the 'sit flask' situation (Schantz *et al.*, 1988), the lower aqueous layer is drawn off through a side arm. Insertion of a pipette through the octanol layer to the aqueous may upset the equilibrium concentrations.

Ideally, both phases should be analysed. This requirement may be circumvented, for solutes of $\log K_{ow}$ not too different from zero, by a knowledge of the concentration in one solvent both before and after partitioning, and of the volumes of the phases. For solutes of very high or very low K_{ow}, one solvent becomes extremely dilute. Under these conditions, the volumes of the phases used are very disparate, and interference from small amounts of solute adsorbed on vessel surfaces becomes significant. If the solute is appreciably volatile, both phases must be analysed. The two analytical methods most often used are absorption spectrophotometry (UV or visible) and high pressure liquid chromatography (HPLC). Among other methods one finds (Sangster, 1993):

gas-(liquid) chromatography
chemical reaction
radiochemical methods
acid-base or coulometric titration
fluorescence
electron spin resonance
nuclear magnetic resonance
gravimetry
atomic absorption
enzymatic/microbial analysis
fluorimetry
polarography

If a solute degrades in solution due to oxidation or reaction with the solvent, the classical shake-flask method cannot be used. Partition coefficients of unstable compounds have, however, been measured by time-dependent methods (Bryon *et al.*, 1980; Tomlinson *et al.*, 1980).

1.2 AKUFVE (SWEDISH ACRONYM)

This is an automated version of the shake-flask method (Davis *et al.*, 1976; James *et al.*, 1981). The principle is shown schematically in Figure 3.1. The two phases are agitated together in a mixing chamber and a sample stream of the heterogeneous mixture is withdrawn. The phases are separated by a centrifuge and each is analysed on-line for solute. The separated streams are returned to the mixing chamber. With this apparatus, it is a relatively easy and rapid matter to change temperature, pH, ionic strength, etc. and to observe the effects of different variables.

1.3 RAPID MIX/FILTER PROBE

This is another automated version of the shake-flask method, and has a formal similarity to AKUFVE. Unlike AKUFVE, which separates the phases for

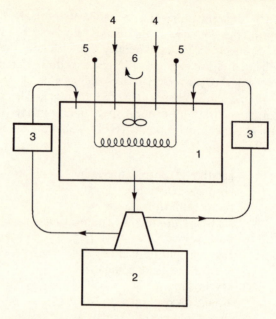

Figure 3.1 Principle of the AKUFVE method: 1, mixing chamber; 2, centrifuge; 3, analysis; 4, solution feeds; 5, temperature control; 6, stirrer (Davis *et al.*, 1976)

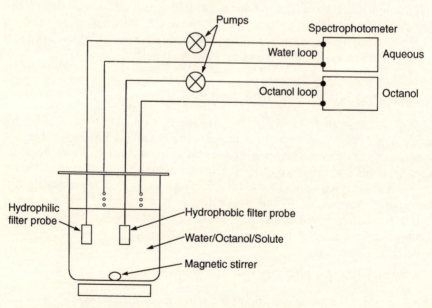

Figure 3.2 Rapid mix/filter probe apparatus (Kinkel *et al.*, 1981)

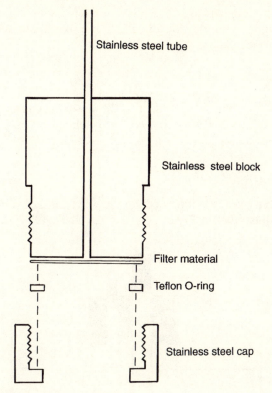

Stainless steel tube

Stainless steel block

Filter material

Teflon O-ring

Stainless steel cap

Figure 3.3 Improved filter probe (Tomlinson, 1982)

analysis by centrifugation, rapid mix/filter probe effects the separation by means of hydrophilic and hydrophobic filters ('filter probe'). The method is shown schematically in Figure 3.2 (Kinkel *et al.*, 1981; Tomlinson *et al.*, 1986). The success of this method is, to a great extent, due to the development of a reliable design for the filter probe. The original construction (Cantwell and Mohammed, 1979) was improved by Tomlinson (1982), Figure 3.3. In the case of the octanol–water system, it has been found more convenient to include only a hydrophilic probe (i.e., to analyse only the aqueous phase). Such a device, slightly modified, has been called a 'filter chamber' (Hersey *et al.*, 1989). In the filter chamber method, the aqueous phase is analyzed as the volume of the octanol phase is changed incrementally. This conveniently accomplishes several determinations in one experiment, without, however, changing the principle of the method.

A closely related device, given the name SEGSPLIT by Tomlinson (1982), uses segmented flow to equilibrate the phases and a phase splitter to separate octanol and water phases (Danielsson and Zhang, 1994; Kinkel and Tomlinson, 1980). In segmented flow, one phase is dispersed as tiny droplets in a stream of the other phase. The phase splitter is simply a small chamber where the liquids separate by dissimilar wetting and hydrophilic and hydrophobic surfaces.

1.4 SIT-FLASK

As mentioned above, this is equivalent to the shake-flask method without the shaking (Schantz *et al.*, 1988). It necessarily eliminates any possibility of emulsion formation.

1.5 SLOW-STIRRING

This can be regarded as intermediate between sit-flask and shake-flask methods. It is illustrated in Figure 3.4 (Brooke *et al.*, 1986). A slightly more

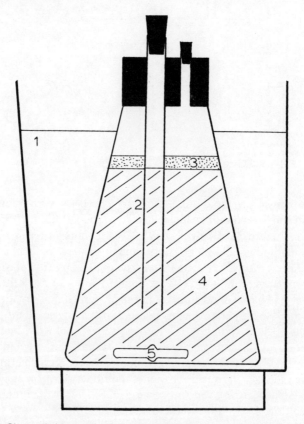

Figure 3.4 Slow-stirring method: 1, constant temperature bath; 2, sampling tube; 3, octanol; 4, water; 5, magnetic stirrer (Brooke *et al.*, 1986)

elaborate design is given by Dearden and Bresnen (1988). In this case the liquid is stirred gently (magnetic stirrer) and the flask is thermostatted. No emulsion forms and the phases are not dispersed. Up to seven days are required for equilibration, depending upon the solute (Brooke et al., 1990; De Bruijn et al., 1989; van Haelst et al., 1994; Sijm et al., 1989; Simpson et al., 1995). Both layers may be sampled with a minimum of disturbance to the equilibrated phases. This method was designed and is particularly useful for very hydrophobic compounds (log $K_{ow} > 5$).

1.6 GENERATOR COLUMN

This is a new design for what is still a direct method, in which the phases are equilibrated without shaking or danger of emulsion formation. The solute is dissolved in octanol, which is pulled through and coats a chromatographic column (~100 mesh Chromosorb W was first used: DeVoe et al., 1981; Wasik et al., 1981). The solute is then eluted with water. In this way, there is a large area of contact for equilibration with no shaking. Since the apparatus is a flow system, solute is continuously depleted from the octanol solution on the solid support; sampling and analysis become somewhat more complicated. Analysis of the aqueous phase is usually by gas chromatography (preceded by solvent extraction) or HPLC. The method was used by the NBS group principally for various types of compounds of log K_{ow} between 0.4 and 5.5 (Schantz and Martire, 1987; Tewari et al., 1982b; Wasik et al., 1981). Because the solute is depleted during measurement, the method is better suited for very hydrophobic compounds and has been employed for PCBs up to log $K_{ow} = 8.2$ (Hawker and Connell, 1988; Larsen et al., 1992; Miller et al., 1984; Woodburn et al., 1984).

1.7 HEAD-SPACE CHROMATOGRAPHIC METHOD

Despite the name, this is a quick-and-dirty version of the shake-flask method, though the phases are not analysed directly (Dallas and Carr, 1992; Hutchinson et al., 1980). In this method the vapour mixture above an aqueous solution (volume v_1) of the solute is sampled by gas chromatography and a chromatographic peak height (h_1) for the solute is obtained. A volume v_2 of octanol is added, and after equilibration, the vapour is sampled and analysed as before (h_2). With certain assumptions, it may be derived that

$$K_{ow} = v_1(h_1/h_2)/v_2 \qquad (3.1)$$

The accuracy of the method is limited by the various assumptions made as well as by the relative precision of v_1/v_2 and h_1/h_2. Its use is necessarily restricted to compounds of sufficient vapour pressure at temperature of measurement.

2 INDIRECT METHODS

Among these are a number of methods based upon correlation of capacity factors of chromatography. Some are very widely used, and the relevant literature is enormous. No attempt has been made in this survey for absolute bibliographic completeness; all salient features are, however, at least mentioned.

2.1 LIQUID CHROMATOGRAPHY WITH SOLID SUPPORT

The very common method of separating substances in solution by passage through a packed column has been used for a long time indeed. The principles are set forth by Snyder and Kirkland (1979). Adaptation to measurement of K_{ow} is summarized in reviews (Braumann, 1986; Hafkenscheid and Tomlinson, 1986; Lambert, 1993; Terada, 1986) from both theoretical and practical viewpoints. The liquid chromatographic method may be realized in the form of tubular columns or thin layers on flat supports. The more widely used columnar version is described first.

As used currently for K_{ow} measurement, the column is packed with finely divided silica gel (3–70 μm) in which hydrocarbon groups are covalently bonded to the silanol sites. Although octadecyl silica (ODS) is most common, octyl or phenyl end groups have also been used. The eluent is a solvent/water (solvent/buffer) mixture of various proportions, common solvents being methanol, acetonitrile, tetrahydrofuran or acetone. Thus a polar solvent elutes the solutes from a non-polar substrate, constituting a 'reversed-phase' and leading to the name *reversed-phase high performance liquid chromatography* (RP-HPLC).

In HPLC, the affinity of a solute for the stationary phase is characterized by the capacity factor k'

$$k' = (t_R - t_0)t_0 \tag{3.2}$$

where t_R is the solute retention time and t_0 is the mobile phase hold-up time. The determination of t_0 in RP-HPLC is generally given by the retention time of a 'non-retained' marker. Often this is an inorganic salt, although a dilute solution of water in the mobile phase is appropriate (Hafkenscheid and Tomlinson, 1986). Since retention volume is directly proportional to retention time at a given flowrate, there is a volume expression equivalent to Eq. (3.2).

In thermodynamic terms, chromatographic retention may be regarded as an equilibrium partition process occurring between two 'immiscible' phases. If the mole fraction equilibrium constant for this process is K_x, then

$$k' = K_x(V_s/V_m) \tag{3.3}$$

where V_s and V_m are volumes occupied by the stationary and mobile phases respectively. Eq. (3.3) has a formal resemblance to Eq. (2.60). The form of Eq. (3.3) suggests that there might be a relation between k' and K_{ow} such as

$$\log K_{ow} = a \log k' + b \tag{3.4}$$

which will be recognized as a Collander-type correlation. Here a and b are empirical constants. Hafkenscheid and Tomlinson (1986) and Braumann (1986) summarize a large number of such correlations reported in 1986.

For a given solute and stationary phase, k' will in general depend on the composition of the mixed solvent used in elution. Experimentally, it was found that $\log k'$ is approximately linearly related to the volume fraction, φ, of organic co-solvent:

$$\log k' = \log k' + c\,\varphi \tag{3.5}$$

where k_w is the capacity factor for pure water as eluent. The $\log k'$–φ curves for many compounds cross, i.e., the rank order of the apparent partition coefficient changes significantly with φ. Some have claimed that only k_w is the reliable measure of K_{ow}. Since k_w is very often difficult, if not impossible, to determine directly (due to inordinately long retention times and spreading of the chromatographic peaks), it is common practice to measure k' at a few compositions and extrapolate to zero co-solvent concentration. In actuality, Eq. (3.5) should be written more correctly as (Miller and Poole, 1994).

$$log\,k' = log\,k_w + c\,\varphi + d\,\varphi^2 \tag{3.6}$$

The question 'to extrapolate or not to extrapolate' (Lambert, 1993) is a matter of some debate. For methanol/water eluents, there is little or no detectable curvature. This is not necessarily true for the other co-solvents.

Partly to bypass these considerations, the idea of making octanol itself the stationary phase has been tried, with apparent success, by groups associated with Taylor (Lewis *et al.*, 1983a,b; Mirrlees *et al.*, 1976), Unger (Henry *et al.*, 1976; Unger *et al.*, 1978, 1986) and others (Caron and Shroot, 1984; Chen *et al.*, 1993). In these cases, octanol (or oleyl alcohol) is coated onto ODS or diatomaceous earth. Coating the column with octanol may be done by passing the liquid through the packing. It may also be accomplished by flowing octanol-saturated water through. The column thus only gradually becomes coated; this is a slow process, since the amount of octanol dissolved in water is very low. Once the column is coated in this way, chromatography is performed as usual and has been given the name *solvent-generated liquid–liquid chromatography* (Cichna *et al.*, 1995). This procedure is claimed to eliminate column 'bleeding' (loss of stationary octanol phase) which occurs when the coating is done directly.

Much the same type of thinking lies behind the use of glycerol-coated controlled-pore glass beads (Miyake and Terada, 1982; Miyake *et al.*, 1988). A silica-type packing, treated so as to present diol groups (—$CH(OH)CH_2OH$) to the eluent, is suggested to have a similar effect (Roumeliotis and Unger, 1979; Terada, 1986).

Since silica-based supports are attacked by high pH solvents (useful for compounds whose pK_as are greater than 7), certain polymers have been tried which are stable to alkali. Packings made from styrene-divinylbenzene copolymer have most often been employed, both as is (Bechalaney et al., 1989; De Biasi et al., 1986; Miyake et al., 1987; Nakae and Muto, 1976; Nakae and Kurihiro, 1978) or C_{18}-derivatized (Lambert and Wright, 1989). Poly(octadecylethylene) was also reported (Bechalaney et al., 1989) as well as triple copolymers (Matsuda et al., 1979).

It is customary to perform elutions in RP-HPLC either with an isocratic (single composition) or polycratic (multiple composition) solvent. There is a further option, namely, gradient elution, in which the composition is programmed to be changed during the elution step. The composition of the mobile phase may be changed linearly or exponentially with time (Kaune et al., 1995; Klamer and Beekman, 1995; Makovskaya et al., 1995). This enables compounds of both short and long retention times to be run in a single experiment.

Thin-layer chromatography (TLC) may conveniently be thought of as a two-dimensional version of columnar liquid chromatography. An informative overview is presented by Zlatkis and Kaiser (1977) and its application in K_{ow} measurement is set forth in three reviews (Biagi et al., 1991; Kaliszan, 1981; Tomlinson, 1975). In early work, the technique was paper chromatography, which uses filter paper as is or impregnated with, e.g., a formamide–ammonium formate mixture. The paper serves as a packed column. A drop of solution containing the sample is introduced at one end of the paper, which is dipped into a liquid mobile phase. The mobile phase ascends (or descends) the length of the paper by capillary action, simultaneously effecting a partition process between the stationary and mobile phases. After the solvent has traversed the length of the strip, the paper is removed and dried. The distance travelled by the spot (L_{solute}) is noted together with the distance travelled by the solvent ($L_{solvent}$) and the quantity

$$R_F = L_{solute}/L_{solvent} \tag{3.7}$$

is called the *retardation factor*. If the spot is not already coloured, it may be made visible by various means (chemical reaction by spray, fluorescence, etc.).

In its modern version, the paper is replaced by a thin layer of powdered adsorbent material on a glass plate. The adsorbent is silica gel, alumina, cellulose, diatomaceous earth, polymer, etc, and is sometimes impregnated or coated with silicone oil, oleyl alcohol, paraffin or octanol. As in RP-HPLC, the adsorbent may also be ODS. The TLC equivalent of the capacity factor in columnar liquid chromatography is

$$R_M = \log(1/R_F - 1) \tag{3.8}$$

which is then correlated with log K_{ow}:

$$\log K_{ow} = a R_M + b \tag{3.9}$$

As in RP-HPLC, the mobile phase in TLC is commonly an aqueous mixture of methanol, acetonitrile or acetone and the question 'to extrapolate or not to extrapolate' also applies.

In the HPLC and TLC methods just described, and in other correlative chromatographic methods discussed below, a calibration curve must be established before K_{ow} values can be deduced. The choice of compounds for this calibration step is not a trivial matter. The success or otherwise of these methods depends, sometimes crucially, on this choice. Further discussion is given in Chapter 4.

Gas-liquid chromatography (GLC) has been used in a manner similar to HPLC and TLC to estimate K_{ow}. In the technique elaborated by Papp, Valko and co-workers, the retention indices of compounds on two liquid stationary phases are correlated with log K_{ow} (Papp *et al.*, 1982; Valko *et al.*, 1984). Various liquid phases were used in the temperature range 100–200°C, including Carbowax (Papp *et al.*, 1983). The closely related *capillary gas chromatography* (Ballschmiter and Zell, 1980) functions in a similar way. Liquid phases include silicone oil, Apiezon L, Apolane 87 (both high molecular weight hydrocarbons) particularly suited for PCBs as solutes (Ballschmiter and Zell, 1980; Risby *et al.*, 1990).

2.2 LIQUID CHROMATOGRAPHY WITHOUT SOLID SUPPORT

Countercurrent chromatography (CCC) is a development of countercurrent extraction (separation) originally described by Craig (1950). This was an automated device for carrying out a series of partitioning steps in a large number of cells. In CCC, the number of steps is greatly increased, the device is much more compact and rugged and the operation is speeded up. The determination of partition coefficients is one application of CCC. The literature on the theory and practice of CCC is voluminous and confusing to the uninitiated. Perhaps the clearest expositions of this method are given by Berthod (1995) and Foucault (1991).

CCC has been carried out in many different kinds of apparatus (Foucault, 1991). Table 3.1 lists some of these. It is probably easiest to grasp the general principle of CCC through the droplet variety (Tanimua *et al.*, 1970). This is shown in Figure 3.5. In this example, the mobile (lighter) phase passes through the heavier stationary phase in the form of droplets and equilibration is achieved before the exit tube is reached. In CCC, the working relationship is (Berthod, 1995)

Table 3.1 Types of countercurrent chromatography (Menges, 1993)

Coiled planet centrifugal
Hydrostatic equilibrium
Hydrodynamic equilibrium
Helix countercurrent
Toroidal coil centrifugal
Droplet
Rotation locular
High-speed
Centrifugal partition chromatography

$$V_R = V_M + K V_S = V_T + (K - 1) V_S \qquad (3.10)$$

where V is volume and M, R, S and T indicate mobile phase, retention, stationary phase and total ($V_T = V_M + V_S$). K is the partition coefficient. This equation applies to all the devices listed in Table 3.1.

The particular method known as *centrifugal partition chromatography* (CPC) is often used for determining K_{ow} (Gluck *et al.*, 1995; Tsai *et al.*, 1995). CPC is similar to droplet CCC (Figure 3.5) except that a centrifugal force, rather than gravitational force, retains the liquid stationary phase. The original design (Murayama *et al.*, 1982) served as the basis of commercially made instruments. In CPC, the column is replaced by a series of cartridges. The cartridges contain a large number of channels created from depressions in blocks of poly(chlorotrifluoroethylene) separated by Teflon sheets. Berthod and Armstrong (1988) describe the general features of CPC.

The use of CPC for the determination of K_{ow} has been reviewed by Gluck and Martin (1990) and also by El Tayar *et al.* (1991). CPC proponents distinguish between 'direct' and 'indirect' modes of operation. (It should be borne in mind that this use of 'direct' and 'indirect' is quite distinct from the present author's use of these terms. As mentioned at the beginning of this chapter, the author's definition of these terms refers to quantitative analysis of the phases. In CPC, there is no quantitative analysis.) In 'direct' CPC, V_M and V_S in Eq. (3.10) are determined by a calibration run with compounds of known K_{ow}; the phases are octanol and water. The retention volume V_R is then measured for each solute. In 'direct' CPC, partition coefficients are determined for another solvent pair (such as hexane-acetonitrile) and a correlation plot of the usual kind is set up. It is not always clear—even to the experts—whether the reported mode is 'direct' or 'indirect' (Gluck and Martin, 1990).

Micellar electrokinetic capillary chromatography is a technique which only recently has been applied to K_{ow} measurement. It can best be grasped through

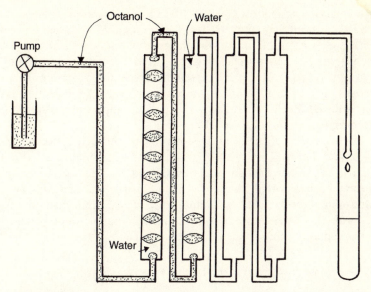

Figure 3.5 Droplet countercurrent chromatography (Menges, 1993)

a detour. RP-HPLC can be modified by changing the mobile phase to an aqueous micellar solution. In this arrangement, the solute partitions itself among three media pairs: water/stationary phase, water/micelles/ and micelles/ stationary phase (Armstrong and Nome, 1981). Imagine now that the stationary phase is removed, and a high dc voltage is applied across the length of the column. There is no solid support, and hence the system may be considered as comprising two immiscible liquid phases. The solute distributes itself between these two phases. Under the electrical field the micelles (usually of sodium dodecylsulphate or cetyltrimethyl ammonium bromide), being charged, migrate at a certain velocity (electrophoresis). The neutral solute dissolved in the water also migrates. It was found that the optimum conditions for separation are

> Column length: 50–90 cm
> Internal diameter: 0.05 mm
> Electric field: $\sim$300 V/cm
> Current: $\leqslant$ 100 μA.

It is necessary also to have efficient temperature control. Under these conditions, the usual liquid chromatographic capacity factor is

$$k' = (t_R - t_0)/t_0(1 - t_R/t_{MC}) \tag{3.11}$$

where t is retention time and R, 0 and MC refer respectively to the solute, a non-retained substance and the micelles (Terabe et $al.$, 1984). The non-retained substance (e.g., methanol) is not solubilized by the micelles; a substance completely solubilized by micelles (e.g., Sudan III) is used to determine t_{MC}. The term $(1 - t_R/t_{MC})$ in Eq. (3.11) comes from the retention behaviour characteristic of electrokinetic separations. When t_{MC} becomes infinite (stationary phase), Eq. (3.11) reduces to Eq. (3.2), applicable in conventional chromatography.

A linear correlation of the usual type may be established between log k' of Eq. (3.11) and log K_{ow} (Greenaway et $al.$, 1994; Ishihama et $al.$, 1994; Muijselaar et $al.$, 1994; Takeda et $al.$, 1993) and K_{ow} data obtained by this method have been reported (Adlard et $al.$, 1995; Herbert and Dorsey, 1995; Smith and Vinjamoori, 1995). Ishihama et $al.$ (1995) replaced the micellar solution with a microemulsion—made from sodium dodecylsulphase + 1-butanol + heptane—and claimed superior results.

2.3 K_{OW} FROM ELECTROMETRIC TITRATION

When a compound is ionizable in aqueous solution (e.g. organic acids and bases), a type of potentiometric titration may be used to determine K_{ow}. This method has been applied with widely varying degrees of sophistication.

The principle is perhaps most easily understood through the procedure of Brändström (1963). An acid or base is titrated in aqueous solution until a fraction of the un-ionized compound is generated and the pH is noted. Then a quantity of octanol is added. Some of the neutral compound passes into the octanol layer, causing the pH in the aqueous phase to change. Further titrant is added in order to restore the pH to its original value. From a knowledge of the quantity of compound and volumes of solvents and titrant, the value of K_{ow} for the neutral compound may be calculated. The pertinent equations are given by Brändström as well as others (Bird and Marshall, 1971; Lassiani et $al.$, 1989; Le Therizien et $al.$, 1980). Multiple titration data points in the presence and absence of octanol can be used to solve equations simultaneously for K_{ow} and for the pK_a of the compound (Seiler, 1974). These equations can be complicated and may be solved by computer (Hersey et $al.$, 1989).

The essence of the method can be stated quite simply: one determines the pK_a of the compound in water, and then measures an apparent value, pK_a', in the presence of octanol. The difference is ΔpK_a and the governing relation is

$$K_{ow} = V_W \left[\text{antilog} \left(\Delta pK_a \right) - 1 \right] / V_{oct} \tag{3.12}$$

where V_W and V_{oct} are the volumes of the two phases at the apparent equivalence point. Thus two titrations are performed (Kaufman et $al.$, 1975) and the entire operation of titration and calculations can be automated (Clarke, 1984). The course of the titration is conventionally represented by a pH–titrant volume plot; however, it is more directly enlightening to subtract a blank titration curve (no sample present). Such a 'difference plot' (Clarke, 1984; Kaufman et al., 1975) is given in Figure 3.6, showing the shift of pK_a upon addition of octanol. Clarke (1984) includes a detailed derivation of the final equations used. In this treatment, it is assumed that (a) only un-ionized species enter the octanol phase, (b) activities may be replaced by concentrations and (c) the volume of the aqueous phase remains approximately constant during the titration. This method was used to determine pK_a and K_{ow} of propanolol esters (Jordan et al., 1992); Quigley et al., 1994) and p-alkoxyphenols (Avdeef, 1991).

The advantages in using difference plots in electrometric titrations were demonstrated by Clarke and Cahoon (1987) and Avdeef (1992). These include the ability to measure pK_a of multiprotic and slightly soluble substances. Avdeef, in particular, refined the method and extended it to the study of mellitic acid (benzenehexacarboxylic acid) and niflumic acid (Avdeef, 1993; Takács-Novák et al., 1994). Measurement of ion-pair partition coefficients and extraction constants was also possible, as well as ease of measurement with very difficult soluble compounds such as the prostaglandins and hexachloro-phene (Avdeef et al., 1995; Avdeef and Comer, 1993; Slater et al., 1994). The

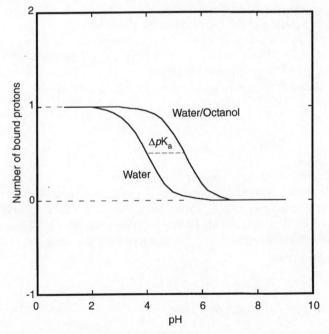

Figure 3.6 Electrometric titration difference plot (Avdeef, 1993)

pK_a of compounds of very low water solubility were determined from Yasudo–Shedlovsky plots for methanol-water solvents, and extrapolated to zero methanol (Avdeef *et al.*, 1993).

2.4 K_{ow} FROM ACTIVITY COEFFICIENTS

It was shown in Chapter 2 that a knowledge of the Henrian activity coefficients of a solute in two separate solvents is equivalent to a knowledge of the partition coefficient (Eq. 2.42). This principle has direct practical usefulness (Sheehan and Langer, 1971; Poe *et al.*, 1993). In application to K_{ow}, activity coefficients of volatile liquids in octanol may be obtained from vapour pressure measurements (Berti *et al.*, 1986) or GLC (Landau *et al.*, 1991; Wasik *et al.*, 1982). For volatiles in water, vapour pressure is used (Pividal *et al.*, 1992) although the most common method is through solubility (Eq. 2.44). For solids, solubility is the source (Kristl and Vesnaver, 1995).

The K_{ow} of various C_1–C_9 compounds were determined from activity coefficients, and accurate results were claimed (Bhatia and Sandler, 1995; Schantz and Martire, 1987; Tse and Sandler, 1994; Wasik *et al.*, 1981).

A related hybrid method may be described here, although it has not yet been applied to K_{ow} (Silveston and Kronberg, 1994). Ordinary liquid chromatography is performed, but in this case the stationary phase is a non-polar liquid polymer (polydimethylsiloxane) coated on non-porous glass beads. The volume of this stationary phase is accurately known. The mobile phase is a dilute solution of the solute in water. The mobile phase is a dilute solution of the solute in water. The partition coefficient of the solute between water and polymer (K_{wp}) may be found directly from the volumes of mobile and stationary phases. The experiment is repeated using a dilute solution of solute in octanol (K_{op}). Then

$$K_{ow} = K_{op}/K_{wp} \tag{3.13}$$

i.e., the polymer phase drops out of consideration.

Another indirect method for obtaining K_{ow} may be extracted from octanol–air partition coefficients (K_{oa}). These have been measured for chlorinated benzenes and PCBs (Harner and Bidleman, 1996; Harner and Mackay, 1995; Harner *et al.*, 1995) by a generator column type method. Now K_{oa} is defined in the same way as K_{ow}:

$$K_{oa} = [X]_o/[X]_a \tag{3.14}$$

and so K_{ow} may be obtained from

$$K_{ow} = ([X]_o/[X]_a)([X]_a/[X]_w) \tag{3.15}$$

$$= K_{oa}K_{aw} \tag{3.16}$$

where K_{aw} is the air–water partition coefficient. K_{aw} is related to the Henry's law constant on the molar concentration scale (H_c)

$$K_{aw} = H_c/RT \tag{3.17}$$

and H_c may be estimated from the vapour pressure of pure liquid solute (p^o) and the water solubility $[X]_w^{ss}$:

$$H_c = p^o/[X]_w^{ss} \tag{3.18}$$

Hence

$$K_{ow} = K_{oa}p^o/RT[X]_w^{ss} \tag{3.19}$$

As an example, the relevant data for 4, 4'-PCB are (Harner and Mackay, 1995)

$$K_{oa} = 4.68E7$$

$$p^o = 4.74E\text{-}8 \text{ atm.}$$

$$RT = 24.5 \text{ L} - \text{atm.}$$

$$[X]_w^{ss} = 2.69E\text{-}7$$

where p^o is taken as the vapour pressure of the supercooled liquid (m. pt. 148°C). Thus calculated, $\log K_{ow} = 5.53$; the experimental value is 5.23 (Sangster, 1993).

2.5 K_{OW} FROM THERMOMETRIC TITRATION

In this method, both K_{ow} and the enthapy of transfer of solute from water to octanol are found simultaneously by calorimetric measurements. This has been accomplished in two ways: (1) titration of an aqueous solution of a neutral solute with octanol, and (2) two-phase acid–base titration.

In (1), octanol is added stepwise to an aqueous solution of the solute and the heat effect is registered. The two unknowns, K_{ow} and $\Delta_{tr}H$, are deduced by solving pertinent equations implicitly. All steps are automated (Fujiwara *et al.*, 1991). The partition coefficients of 15 alkanols and and substituted alkanols were found in this way.

In (2), an organic base is titrated with HCl in a two-phase octanol–water system. Again, K_{ow} and $\Delta_{tr}H$ are determined implicitly from a knowledge of the heat effects involved and the volumes of solutions. In this case, a separate experiment must be performed in the absence of octanol to measure the heat of protonation of the base. This procedure was applied to ephedrine and pseudoephedrine (Burgot and Burgot, 1984, 1986).

2.6 K_{ow} FROM KINETICS OF PARTITIONING

The partitioning of solute X between immiscible phases is a dynamic process:

$$X \text{ (in water)} \underset{k_B}{\overset{k_A}{\rightleftharpoons}} X \text{(in octanol)} \tag{3.20}$$

where the ks are first-order rate constants. As a consequence of this equilibrium,

$$K_{ow} = k_A / k_B \tag{3.21}$$

Eq. (3.21) is the basis of a number of reported methods of deducing partition coefficients from kinetic measurements.

One experimental set-up resembles closely that of the slow-stirring method (q.v., above) in which one phase initially contains all the solute and solute concentration is followed spectrophotometrically as a function of time. Integrated rate equations or graphical methods are used to deduce k_A and k_B and hence K_{ow}. Partition coefficients of benzene derivatives, isothiocyanates and 2-furylethylenes were found in this way (Augustin *et al.*, 1987; Baláž *et al.*, 1985; Schelenz *et al.*, 1993).

In a somewhat different version, drops of one phase containing the solute (ephedrine) passed through the other phase in columns of different lengths. After passage, drops were continuously collected for analysis (Brodin, 1974).

In a third variation, the thermometric titration procedure (case 2 above) was adapted to measure time-dependent heat effects. The partition coefficient of ephedrine, as measured, was the same as that reported by other methods (Burgot and Burgot, 1991).

2.7 K_{ow} FROM WATER SOLUBILITY CORRELATION

A thermodynamic justification of this correlation was given in Chapter 2. A closer look at its empirical adequacy is given in Chapter 4.

REFERENCES

Adlard, M., Okafo, G., Meenan, E. and Camillieri, P. (1995) *J. Chem. Soc. Chem. Commun.*, 2241–3.

Armstrong, D. W. and Nome, F. (1981) *Anal. Chem.* **53**, 1662–6.

Augustín, J., Baláž, Š., Hanes, J. and Šturdík, E. (1987) *Chem. Papers* **41**, 401–5.

Avdeef, A. (1991) *Pharmacochem. Libr.* **16**, 119–22.

Avdeef, A. (1992) *Quant. Struct.-Act. Relat.* **11**, 510–17.

Avdeef, A. (1993) *J. Pharm. Sci.* **82**, 183–90.

Avdeef, A. and Comer, J. E. A. (1993) *Proc. 9th Eur. Symp. Struct.-Act. Relat.–QSAR Mod. Modell.*, 386–7.

Avdeef, A., Box, K. J. and Takacs-Novak, K. (1995) *J. Pharm. Sci.* **84**. 523–9.

Avdeef, A., Comer, J. E. A. and Thomson, S. J. (1993) *Anal. Chem.* **65**, 42–9.

Baláž, Š., Kuchár, A., Šturdik, E., Rosenberg, M., Štibrányi, L. and Ilavský, D. (1985) *Coll. Czeh. Chem. Commun.* **50**, 1642–7.

Ballschmiter, K. and Zell, M. (1980) *Fresenius Z. Anal. Chem.* **302**, 20–31.

Bechalaney, A., Rothlisberger, T., El Tayar, N. and Testa, B. (1989) *J. Chromatogr.* **473**, 115–24.

Berthod, A. (1995) *Chromatogr. Sci. Ser.* **68**, 167–97.

Berthod, A. and Armstrong, D. W. (1988) *J. Liq. Chromatogr.* **11**, 547–66.

Berti, P., Cabani, S., Conti, G. and Mollica, V. (1986) *J. Chem. Soc. Faraday Trans I.* **82**, 2547–56.

Bhatia, S. R. and Sandler, S. I. (1995) *J. Chem. Eng. Data.* **40**, 1196–8.

Biagi, G. L., Recanatini, M., Barbaro, A. M., Guerra, M. C., Sapone, A., Borea, P. A. and Pietrogrande, M. C. (1991) *Pharmacochem. Libr.* **16**, 83–90.

Bird, A. E. and Marshall, A. C. (1971) *J. Chromatogr.* **63**, 313–19.

Brändström, A. (1963) *Acta Chem. Scand.* **17**, 1218–24.

Braumann, T. (1986) *J. Chromatogr.* **373**, 191–225.

Brodin, A. (1974) *Acta Pharm. Suec.* **11**, 141–8.

Brooke, D. N., Dobbs, A. J. and Williams, N. (1986) *Ecotoxocol. Environ. Chem.* **11**, 251–60.

Brooke, D., Nielsen, I., de Bruijn, J. and Hermens, J. (1990) *Chemosphere* **21**, 119–33.

Burgot, G. and Burgot, J.-L. (1984) *Thermochimica Acta.* **81**, 147–56.

Burgot, G. and Burgot, J.-L. (1986) *Ann. Pharm. Fr.* **44**, 313–16.

Burgot, G. and Burgot, J.-L. (1991) *Thermochimica Acta.* **180**, 49–59.

Byron, P. R., Notari, R. E. and Tomlinson, E. (1980) *J. Pharm. Sci.* **69**, 527–31.

Cantwell, F. F. and Mohammed, H. Y. (1979) *Anal. Chem.* **51**, 218–23.

Caron, J. C. and Shroot, B. (1984) *J. Pharm. Sci.* **73**, 1703–6.

Chen, C., Bailey, E. J., Hartley, C. D. *et al.*, (1993) *J. Med. Chem.* **35**, 3646–57.

Cichna, M. Markl, P. and Haber, J. F. K. (1995) *J. Pharm. Biomed. Anal.* **13**, 339–51.

Clarke, F. H. (1984) *J. Pharm. Sci.* **73**, 226–30.

Clarke, F. H. and Cahoon, N. M. (1987) *J. Pharm. Sci.* **76**, 611–20.

Craig, L. C. (1950) *Anal. Chem.* **22**, 1346–52.

Dallas, A. J. and Carr, P. W. (1992) *J. Chem. Soc. Perkin Trans.* 2, 2155–61.

Danielsson, L.-G. and Zhang, Y.-H. (1994) *J. Pharm. Biomed. Anal.* **12**, 1475–81.

Davis, S. S., Elson, G., Tomlinson, E., Harrison, G. and Dearden, J. C. (1976) *Chem. Ind. (London)* (16), 677–83.

Dearden, J. C. and Bresnen, G. M. (1988) *Quant. Struct.-Act. Relat.* **7**, 133–44.

De Biasi, V., Lough, W. J. and Evans, M. B. (1986) *J. Chromatogr.* **353**, 279–84.

De Bruijn, J., Busser, F., Seinen, W. and Hermens, J. (1989) *Environ. Toxicol. Chem.* **8**, 499–512.

DeVoe, H., Miller, M. M. and Wasik, S. P. (1981) *J. Res. Nat. Bur. Stand. (U.S.)* **86**, 361–6.

Dunn, W. H., Block, J. H. and Pearlman, R. S., (eds.) (1986) *Partition Coefficient Determination and Estimation*, Pergamon Press, New York.

El Tayar, N., Tsai, R.-S., Vallat, P., Altomare, C. and Testa, B. (1991) *J. Chromatogr.* **556**, 181–94.

Foucault, A. P. (1991) *Anal. Chem.* **63**, 569A–79A.

Fujiwara, H., Yoshikawa, H. Murata, S. and Sasaki, Y. (1991) *Chem. Pharm. Bull.* **39**, 1095–8.

Gluck, S. J. and Martin, E. J. (1990) *J. Liq. Chromatogr.* **13**, 2529–51.

Gluck, S. J., Martin, E. and Benko, M. H. (1995) *Chromatogr. Sci. Ser.* **68**, 199–218.

Greenaway, M., Okafo, G., Manallack, D. and Camilieri, P. (1994) *Electrophoresis* **15**, 1284–9.

van Haelst, A. G., Heesen, P. F., van der Wielen, F. W. M. and Govers, H. A. J. (1994) *Chemosphere* **29**, 1651–60.

Hafkenscheid, T. L. and Tomlinson, E. (1986) *Adv. Chromatogr.* **25**, 1–62.

Hansch, C., (gen. ed.) (1990) *Comprehensive Medicinal Chemistry*, 6 vols., Pergamon Press, New York.

Harner, T. and Bidleman, T. F. (1996) *J. Chem. Eng. Data* **41**, 895–9.

Harner, T. and Mackay, D. (1995) *Environ. Sci. Technol.* **29**, 1599–1606.

Harner, T., Bidleman, T., Falconer, R. and Mackay, D. (1995) *Organohalogen Compd.* **24**, 445–8.

Hawker, D. W. and Connell, D. W. (1988) *Environ. Sci. Technol.* **22**, 382–7.

Henry, D., Block, J. H., Anderson, J. L. and Carlsson, G. R. (1976) *J. Med. Chem.* **19**, 619–26.

Herbert, B. J. and Dorsey, J. G. (1995) *Anal. Chem.* **67**, 744–9.

Hersey, A., Hill, A. P., Hyde, R. M. and Livingstone, D. J. (1989) *Quant. Struct.-Act. Relat.* **8**, 288–96.

Hutchinson, T. C., Hellebust, J. A., Tam, D., Mackay, D., Mascarenhas, R. A. and Shiu, W.-Y. (1980) In B. K. Afghan and D. Mackay, (eds.) (1980) *Hydrocarbons and Halogenated Hydrocarbons in the Aquatic Environment*, Plenum Press, New York, pp. 577–86.

Ishihama, Y., Oda, Y., Uchikawa, K. and Asakawa, N. (1994) *Chem. Pharm. Bull.* **42**, 1525–7.

Ishihama, Y., Oda, Y., Uchikawa, K. and Asakawa, N. (1995) *Anal. Chem.* **67**, 1588–95.

James, M. J., Davis, S. S. and Anderson, N. (1981) *J. Pharm. Pharmacol.* **33**, 108P.

Jordon, C. G. M., Quigley, J. M. and Timoney, R. F. (1992) *Int. J. Pharm.* **84**, 175–89.

Kaliszan, R. (1981) *J. Chromatogr.* **220**, 71–83.

Kaufman, J. J., Semo, N. M. and Koski, W. S. (1975) *J. Med. Chem.* **18**, 647–55.

Kaune, A., Knorrenschild, M. and Kettrup, A. (1995) *Fresenisu' J. Anal. Chem.* **352**, 303–12.

Kinkel, J. F. M. and Tomlinson, E. (1980) *Int. J. Pharm.* **6**, 261–75.

Kinkel, J. F. M., Tomlinson, E. and Smit, P. (1981) *Int. J. Pharm.* **9**, 121–36.

Klamer, H. J. C. and Beekman, M. (1995) *Toxicol. Model.* **1**, 169–79.

Kristl, A. and Vesnaver, G. (1995) *J. Chem. Soc. Faraday Trans.* **91**, 995–8.

Lambert, W. J. (1993) *J. Chromatogr. A* **656**, 469–84.

Lambert, W. J. and Wright, L. A. (1989) *J. Chromatogr.* **464**, 400–4.

Landau, I., Belfer, A. J. and Locke, D. C. (1991) *Ind. Eng. Chem. Res.* **30**, 1900–6.

Larsen, B., Skejø-Andreasen, H. and Paya-Perez, A. (1992) *Fresenius Environ. Bull.* **1**(Suppl.), 513–18.

Lassiani, L., Ebert, C., Nisi, C. and Varnavas, A. (1989) *Farmaco* **44**, 1239–43.

Leo, A. (1991) *Methods in Enzymology* **202**, 544–91.

Leo, A., Hansch, C. and Elkins, D. (1971) *Chem. Rev.* **71**, 525–616.

Le Therizien, L., Heymans, F., Redevilh, C. and Godfroid, J.-J. (1980) *Eur. J. Med Chem. Chim. Ther.* **15**, 311–16.

Lewis, S. J., Mirrlees, M. S. and Taylor, P. J. (1983a) *Quant. Struct.-Act. Relat.* **2**, 1–6.

Lewis, S. J., Mirrlees, M. S. and Taylor, P. J. (1983b) *Quant. Struct.-Act. Relat.* **2**, 100–11.

Makovskaya, V., Dean, J. R., Tomlinson, E., Hitchen, S. M. and Comber, M. (1995) *Anal. Chim. Acta.* **315**, 183–92.

Martin, Y. C. (1978) *Quantitative Drug Design*, Marcel Dekker, New York, Chapter 4.

Matsuda, R., Yamamiya, T., Tatsuzawa, M. Ejima, A. and Takai, N. (1979) *J. Chromatogr.* **173**, 75–8.

Menges, R. A. (1993) *Ph. D. Thesis*, University of Missouri-Rolla.

Miller, K. G. and Poole, C. F. (1994) *J. High Resolut. Chromatogr.* **17**, 125–34.
Miller, M. M., Ghodbane, S., Wasik, S. P., Tewari, Y. B. and Martire, D. E. (1984) *J. Chem. Eng. Data* **29**, 184–90.
Mirrlees, M. S., Moulton, S. J., Murphy, C. T. and Taylor, P. J. (1976) *J. Med Chem.* **19**, 615–19.
Miyake, K. and Terada, H. (1982) *J. Chromatogr.* **240**, 9–20.
Miyake, K., Kitaura, F., Mizuno, N. and Terada, H. (1987) *Chem. Pharm. Bull.* **35**, 377–88.
Miyake, K., Mizuno, N. and Terada, H. (1988) *J. Chromatogr.* **439**, 227–35.
Muijselaar, P. G. H. M., Claessens, H. A. and Cramers, C. A. (1994) *Anal. Chem.* **66**, 635–44.
Murayama, W., Kobayashi, T., Kosuge, Y., Yano, H., Nunogaki, Y. and Nunogaki, K. (1982) *J. Chromatogr.* **239**, 643–9.
Nakae, A. and Kurihiro, K. (1978) *J. Chromatogr.* **156**, 167–72.
Nakae, A. and Muto, G. (1976) *J. Chromatogr.* **120**, 47–54.
Papp, O., Jozan, M., Valko, K., Hankone Novak, K., Hermecz, I. and Szasz, G. (1983) *Acta Pharm. Hung.* **53**, 215–21.
Papp, O., Valko, K., Szasz, G., Hermecz, I., Vamos, J., Hanko, K. and Ignath-Halasz, Z. (1982) *J. Chromatogr.* **252**, 67–75.
Pividal, K. A., Birtigh, A. and Sandler, S. I. (1992) *J. Chem. Eng. Data* **37**, 484–7.
Poe, R. B., Rutan, S. C., Hait, M. J., Eckert, C. A. and Carr, P. W. (1993) *Anal. Chim. Acta* **277**, 223–38.
Purcell, W. P., Bass, G. E. and Clayton, J. M. (1973) *Strategy of Drug Design*, Wiley Interscience, New York, Appendix I.
Quigley, J. M., Jordon, C. G. M. and Timoney, R. F. (1994) *Int. J. Pharm.* **101**, 145–63.
Rekker, R. F. (1977) *The Hydrophobic Fragmental Constant*, Elsevier, Amsterdam, Chap. 1.
Risby, T. H., Hsu, T.-B., Sehnert, S. S. and Bhan, P. (1990) *Environ. Sci. Technol.* **24**, 1680–7.
Roumeliotis, P. and Unger, K. K. (1979) *J. Chromatogr.* **185**, 445–52.
Sangster, J. (1989) *J. Phys. Chem. Ref. Data* **18**, 1111–229.
Sangster, J. (1993) *LOGKOW—a Databank of Evaluated Octanol-water Partition Coefficients*, Sangster Research Laboratories, Montreal.
Schantz, M. M. and Martire, D. E. (1987) *J. Chromatogr.* **391**, 35–51.
Schantz, M. M., Barman, B. N. and Martire, D. E. (1988) *J. Res. Nat. Bur. Stand. (U.S.)* **93**, 161–73.
Schelenz, T., Kramer, C.-R. Henze, U. and Lübeck, K. (1993) *J Prakt. Chem.* **335**, 273–8.
Seiler, P. (1974) *Eur. J. Med. Chem.-Chim. Ther.* **9**, 663–5.
Sheehan, R. J. and Langer, S. H. (1971) *Ind. Eng. Chem. Process Des. Develop.* **10**, 44–7.
Sijm, D. T. H. M., Wever, H., de Vries, P. J. and Opperhuizen, A. (1989) *Chemosphere* **19**, 263–6.
Silveston, R. and Kronberg, B. (1994) *J. Chromatogr. A* **659**, 43–56.
Simpson, C. D., Wilcock, R. J., Smith, T. J., Wilkins, A. L. and Langdon, A. G. (1995) *Bull. Environ. Contam. Toxicol.* **55**, 149–53.
Slater, B., McCormack, A., Avdeef, A. and Comer, J. E. A. (1994) *J. Pharm. Sci.* **83**, 1280–3.
Smith, J. T. and Vinjamoori, D. V. (1995) *J. Chromatogr. B* **669**, 59–66.
Snyder, L. R. and Kirkland, J. J. (1979) *Introduction to Modern Liquid Chromatography*, 2nd edition, John Wiley and Sons, New York.
Takács-Novák, K., Avdeef, A., Box, K. J., Podányi and Szász, G. (1994) *J. Pharm. Biomed. Anal.* **12**, 1369–77.

Takeda, S., Wakida, S.-I., Yamane, M., Kawahara, A. and Higashi, K. (1993) *Anal. Chem.* **65**, 2489–92.

Tanimura, T., Pisaro, J. J., Ito, Y. and Bowman, R. L. (1970) *Science* **169**, 54–6.

Taylor, P. J. (1990) in Volume 4 of Hansch (1990), Chap. 18.6.

Terabe, S., Otsuka, K., Ichikawa, K., Tsuchiya, A. and Ando, T. (1984) *Anal. Chem.* **56**, 111–13.

Terada, H. (1986) *Quant. Struct.-Act. Relat.* **5**, 81–8.

Tewari, Y. B., Martire, D. E., Wasik, S. P. and Miller, M. M. (1982a) *J. Solution Chem.* **11**, 435–45.

Tewari, Y. B., Miller, M. M., Wasik, S. P. and Martire, D. E. (1982b) *J. Chem. Eng. Data* **27**, 451–4.

Tomlinson, E. (1975) *J. Chromatogr.* **113**, 1–45.

Tomlinson, E. (1982) *J. Pharm. Sci.* **71**, 602–4.

Tomlinson, E., Notari, R. E. and Byron, P. R. (1980) *J. Pharm. Sci.* **69**, 655–8.

Tomlinson, E., Davis, S. S., Parr, G. D et al. (1986) in Dunn *et al* (1986), pp. 83–99.

Tsai, R.-S., Carrupt, P.-A. and Testa, B. (1995) *ACS Symp. Ser.* **593**, 143–54.

Tse, G. and Sandler, S. I. (1994) *J. Chem. Eng. Data* **39**, 354–7.

Unger, S. H., Cook, J. R. and Hollenberg, J. S. (1978) *J. Pharm. Sci.* **67**, 1354–7.

Unger, S. H., Cheung, P. S., Chiang, G. H. and Cook, J. R. (1986) in Dunn *et al.*, (1986), pp. 69–81.

Valko, K., Papp, O. and Darvas, F. (1984) *J. Chromatogr.* **301**, 355–64.

Wang, P.-S. and Lien, E. J. (1980) *J. Pharm. Sci.* **69**, 662–8.

Wasik, S. P., Tewari, Y. B. and Miller, M. M. (1982) *J. Res. Nat. Bur. Stand. (U.S.)* **87**, 311–15.

Wasik, S. P., Tewari, Y. B., Miller, M. M. and Martire, D. E. (1981) *Octanol-water Partition Coefficients and Aqueous Solubilities of Organic Compounds*, NBSIR-81-2406, U. S. Department of Commerce, Washington.

Woodburn, K. B., Doucette, W. J. and Andren, A. W. (1984) *Environ. Sci. Technol.* **18**, 457–9.

Zlatkis, A. and Kaiser, R. E., (eds.) (1977) *High Performance Thin-layer Chromatography*, Elsevier, Amsterdam.

CHAPTER 4

Discussion of Measurement Methods

Chapter 3 describes an array of methods which have been used to measure K_{ow}. At first sight, such a variety might well bewilder. There are a number of discussions in the literature, of varying length and detail, which are helpful in general orientation: these are listed in Table 4.1.

There is no best general-purpose method of K_{ow} measurement, equally applicable to all kinds of solutes. This unremarkable conclusion is often belied by an uncritical reading of the voluminous published reports. The discerning reader will, however, not be misled by professional enthusiasms, the 'fine print', unstated assumptions, selection or omission of data, etc. Nevertheless it is clear that, if nothing but the best will do, then an appropriate direct method must be used. This is particularly true if one is creating or elaborating a K_{ow} calculation method (Leo, 1995). For this purpose one needs to be sure that the known or 'training' data are reliable. The many indirect methods described in Chapter 3 depend on standard data for correlation. Also, a direct method is probably best for the determination of Gibbs energies of transfer.

1 WATER SOLUBILITY CORRELATION

In Chapter 2 it was shown that the simple linear relation

$$\log K_{ow} = a \log S_w + b \tag{2.35}$$

was an approximation to the thermodynamically derived equations

$$\log K_{ow} = -\log c^{w,ss} - \log(\gamma^{oct,ss} V_{oct}) \tag{2.49}$$

for neat solvents and

$$\log K_{ow} = -\log c^{w,ss} - \log(\gamma^{*oct,ss} V_{*oct}) + \log(\gamma^{*w,ss}/\gamma^{w,ss}) \tag{2.74}$$

Table 4.1 Discussions on K_{ow} measurement methods

Methods discussed or mentioned	References
SF, RP-HPLC, CPC	Berthod (1995)
RP-HPLC	Braumann (1986)
SF, GC, RP-HPLC	Brooke et al. (1990)
SF, SS, GC, RP-HPLC, RP-TLC	Chessells et al. (1991)
SF, SS, ET, GC, RP-HPLC, CCC, CPC	Danielsson and Zhang (1996)
SF, SS	Dearden and Bresnen (1988)
SF, SS, GC, RP-HPLC	De Bruijn et al. (1989)
SF, RP-HPLC, RP-TLC, water solubility	Eadsforth and Moser (1983)
CPC	El-Tayar et al. (1989)
SF, RP-HPLC, RP-TLC, water solubility	Esser and Moser (1982)
RP-HPLC	Hafkenscheid and Tomlinson (1986)
SF, RP-HPLC	Harnisch et al. (1983)
RP-HPLC, RP-TLC	Kaliszan (1981)
RP-HPLC, CPC	Kaliszan (1990)
RP-HPLC	Klein et al. (1988)
RP-HPLC, RP-TLC	Lambert (1993)
SF	Leo (1991)
RP-TLC	Rekker (1984)
SF, RP-HPLC, RP-TLC	Renberg et al. (1985)
CCC, RP-HPLC, SF	Taylor (1990)
RP-HPLC	Terada (1986)
RP-TLC	Tomlinson (1975)

CCC = countercurrent chromatography
CPC = centrifugal partition chromatography
ET = electrometric titration
GC = generator column
RP-HPLC = reversed phase high performance liquid chromatography
RP-TLC = reversed phase thin layer chromatography
SF = shake-flask
SS = slow stirring

for mutually saturated solvents. Both these equations are applicable to liquid solutes. They are not linear relations, since activity coefficients are solute dependent. (See Chapter 2 for explanation of symbols.)

Considerable effort has been expended to find out how adequate Eq. (2.35) is in real terms, i.e., with experimental data. Table 4.2 lists the results of a number of reports. According to Eqs (2.49) and (2.74), the slope of the plot should be equal to -1; in practice, it is close to, but usually greater than, -1. In one case (Mackay et al., 1980) a slope of -1 was imposed on the correlation, in order to arrive at an estimate for $\gamma^{oct,ss}$. In other studies (Bowman and Sans, 1983; Chiou et al., 1982) an 'ideal K_{ow}' $= -\log c^{w,ss} - \log V_{oct}$ (i.e., activity coefficient $= 1$) was used as a basis of comparison for actual correlations. The confused state of affairs (Isnard and Lambert, 1989) is at least partly due to the fact that different authors sometimes use different data for the same

Table 4.2 Water solubility correlations for organic liquids; $\log K_{ow} = a \log S_w + b$ (S_w in molarity)

Reference	n	a	b	s	r	F	Type of compounds	log K_{ow} range min, max
Banerjee et al. (1980)	17	−0.723	1.13	0.29	−0.935	104	various	1.45, 4.78
Bowman and Sans (1983)	32	−0.885	0.163	0.27	−0.977	616	insecticides	1.47, 5.68
Chiou et al. (1982)	156	−0.747	0.730	—	−0.935	—	various	—
	41	−0.898	0.832	—	−0.967	—	alcohols	—
	18	−0.987	0.513	—	−0.990	—	esters	—
	13	−0.813	0.586	—	−0.980	—	ketones	—
	16	−0.808	−0.201	—	−0.953	—	alkanes	—
	16	−1.004	0.340	—	−0.975	—	aromatics	—
	20	−0.819	0.681	—	−0.928	—	alkyl halides	—
Hansch et al. (1968)	23	−0.766	0.811	0.28	−0.959	241	various[a]	0.16, 2.99
Isnard and Lambert (1989)	165	−0.799	0.787	0.38	−0.957	1782	various	−0.34, 6.00
Leahy (1986)	77	−0.866	0.766	0.23	−0.980	1815	various[b]	−1.35, 5.18
Mackay et al. (1980)	25	−0.760	0.985	0.18	−0.975	450	various[c]	2.13, 4.90
Miller et al. (1985)	48	−0.874	0.671	0.15	−0.874	1749	various	1.41, 5.52
	13	−0.827	0.744	0.12	−0.995	1055	halogenated hydrocarbons	2.67, 5.47
	12	−0.942	0.558	0.05	−0.999	3654	aromatic hydrocarbons	2.13, 5.52
Valvani et al. (1981)	44	−0.919	0.577	0.38	−0.919	230	various	−0.34, 3.38
Wasik et al. (1981)	28	−0.891	0.730	0.11	−0.997	3108	various	0.68, 5.18

Underlined statistical data were calculated by the author (J.S.).
[a] Only experimental K_{ow} data used
[b] Excluding compounds completely miscible with water and including 19 compounds with Recommended K_{ow} data (Sangster, 1993)
[c] Excluding pesticides
n = number of data points
s = standard deviation or standard error of estimate
r = correlation coefficient
F = F-test (various indices)

compounds in their correlations. Table 4.3 presents a list of some representative compounds and the data (log K_{ow}, log S_w) used in correlations. While there is substantial agreement for many substances, there are also significant discrepancies. Finally, Isnard and Lambert (1989) state that '. . . it is of little importance whether the solubility is expressed in g/m^3 or in mol/m^3; the correlation coefficients and the standard deviations are not, statistically, different . . .'. Expressing S_w in weight rather than mole units effectively puts the term log MW (MW = molecular weight) into the intercept of Eq. (2.35).

Alone among the investigators listed in Table 4.2, the NBS group measured both K_{ow} and water solubility of all their reported compounds by the generator column method. Their data appear in Miller *et al.* (1985), Tewari *et al.* (1982) and Wasik *et al.* (1981). The results of their work underline the importance of using good quality data for correlations. In addition, they illustrated a related correlation

$$\log K_{ow} = a \log f^{w,ss} = b \tag{4.1}$$

where K_{ow} is the measured partition coefficient and $f^{w,ss}$ is the volume fraction Henrian activity coefficient, derived from their water solubility measurements. The statistical data corresponding to Eq. (4.1) were

$$n = 62 \text{ liquids (various)}$$
$$a = 0.944$$
$$b = -0.311$$
$$s = 0.20$$
$$r = 0.990$$
$$\text{Range of } \log K_{ow} \text{ data: } 0.52\text{--}5.35$$

The thermodynamic equivalent of Eq. (4.1) is easily derived. Eq. (2.67) gives the definition of K_{ow} for the case of mutually saturated solvents:

$$K_{ow} = f^{*w}/f^{*oct} \tag{2.67}$$

$$\therefore \log K_{ow} = f^{*w} - \log f^{*oct} \tag{4.2}$$

$$= \log(f^{*w}f^{w,ss}/f^{w,ss}) - \log f^{*oct} \tag{4.3}$$

$$= \log f^{w,ss} + \{\log(f^{*w}/f^{w,ss}) - \log f^{*oct}\} \tag{4.4}$$

where, as before, the asterisk refers to mutually saturated solvents and ss refers to solvent saturated with respect to solute. Eq. (4.4) is not a linear relationship between log K_{ow} and log $f^{w,ss}$. Comparing Eqs (4.4) and (4.1), we see that in the thermodynamic derivation, $a = 1$ and b is equal to the expression in braces. The most significant term in the intercept expression is f^{*oct}. Henrian activity

Table 4.3 Log K_{ow} and log S_w (molarity) data used in published correlations*

Compounds	Banerjee et al. (1980)	Hansch et al. (1968)	Isnard and Lambert (1989)	Leahy (1986)	Mackay et al. (1980)	Wasik et al. (1981)	Valvani et al. (1981)
Hydrocarbons							
Pentane	2.50	2.50	3.39			3.62	
	−3.27	−3.27	−3.27			−3.25	
Hexane			4.11	3.90		4.11	
			−3.85	−3.95		−3.84	
Heptane			4.66	4.66		4.66	
			−4.51	−4.53		−4.45	
Octane			5.18	5.18		5.18	
			−5.23	−5.24		−5.02	
Cyclohexane			3.44	3.44			3.44
			−3.19	−3.18			−3.07
1-Hexene			3.39			3.47	
			−3.08			−3.08	
1-Pentyne			2.12			2.12	
			−1.93			−1.81	
Benzene	2.12	2.13	2.12		2.13		
	−1.65	−1.64	−1.63		−1.64		
Toluene	2.21	2.69	2.63	2.69	2.69	2.65	2.58
	−1.77	−2.29	−2.24	−2.25	−2.25	−2.20	−2.24
Ethylbenzene			3.13			3.13	3.15
			−2.80			−2.75	−2.81
Styrene	3.16		3.16				2.95
	−2.81		−2.54				−2.54
1,3,5-Tri methylbenzene			3.42	3.42			3.42
			−3.08	−3.24			−3.24
1,2,3-Tri methylbenzene			3.55	3.55			
			−3.26	−3.26			

Table 4.3 *(Continued)*

Compounds	Banerjee et al. (1980)	Hansch et al. (1968)	Isnard and Lambert (1989)	Leahy (1986)	Mackay et al. (1980)	Wasik et al. (1981)	Valvani et al. (1981)
Alcohols							
1-Butanol			0.88			0.79	0.75
			−0.01			−0.07	−0.01
2-Butanol		0.61	0.81				
		0.29	0.39				
1-Pentanol			1.40	1.40		1.53	1.48
			−0.60	−0.61		−0.88	−0.60
1-Hexanol			1.99	2.03		2.03	
			−1.24	−1.24		−1.38	
Cyclohexanol		1.23	1.23	1.23			1.23
		−0.42	−0.42	−0.45			−0.30
Ketones							
2-Butanone		0.29	0.35	0.29		0.69	
		0.68	0.52	0.49		0.28	
2-Pentanone			0.99	0.91		0.99	
			−0.27	−0.18		−0.28	
Acetophenone			1.63	1.58			1.66
			−1.34	−1.31			−1.31
Cyclohexanone			0.81	0.81			
			−0.63	0.01			
Ethers							
Diethyl ether			0.83	0.89			0.83
			−0.06	−0.13			−0.06
Furan			1.34				1.34
			−0.83				−0.83
Esters							
Ethyl acetate		0.73	0.73	0.73		0.68	0.70
		−0.04	−0.13	−0.05		−0.14	−0.06

Compound							
Butyl acetate		1.82 / -1.24		1.82 / -1.36	1.73 / -1.24		
Halogenated							
Chloroform	1.96 / -1.12		1.97 / -1.18	1.94 / -1.12	1.90 / -1.17	1.97 / -0.92	1.90 / -1.22
Tetrachloro ethylene			2.60 / -2.62		2.88 / -2.92		2.53 / -2.53
Trichloro ethylene		2.53 / -2.98		2.29 / -1.95	2.53 / -1.98		
1-Chlorobutane		2.55 / -2.03		2.64 / -2.14	2.55 / -2.12		
Chlorobenzene	2.84 / -2.26	2.81 / -2.58	2.84 / -2.38	2.84 / -2.35	2.98 / -2.58	2.84 / -2.36	
Bromobenzene		2.98 / -2.58		2.99 / -2.64	3.02 / -2.54		
Amines							
Aniline	0.93 / -0.15				0.98 / -0.41		
Triethylamine	1.45 / -0.83			1.45 / -0.26	1.45 / -0.83		
Nitro							
Nitrobenzene	1.87 / -2.07	1.85 / -1.51		1.85 / -2.07	1.85 / -1.81	1.85 / -1.78	1.83 / -1.77

*In each pair of data, the first is log K_{ow} and the second, log S_w

coefficients of solutes have been measured in anhydrous octanol by GLC (Schantz and Martire, 1987), by head space chromatography (Dallas and Carr, 1992) and by vapour pressure measurements (Berti et al., 1986; Cabani et al., 1991). Activity coefficients in water-saturated octanol were measured by head-space chromatography (Dallas and Carr, 1992). From these reports, both f^{*oct} and f^{oct} vary significantly from solute to solute, even within homologous series (although they are not as sensitive to chain length or polarity as the corresponding activity coefficients in water).

1.1 THE MELTING POINT CORRECTION

It was seen in Chapter 2 that the thermodynamic relationship between log K_{ow} and log $c^{w,ss}$ for solids was

$$\log K_{ow} = -\Delta_{fus}G_T^o/2.303RT - \log c^{w,ss} - \log(Vf^{oct,ss}) \qquad (2.55)$$

where the first term on the RHS has come to be known at the 'melting point correction'. This term is often approximated by assuming that the heat capacity change upon melting is zero ($\Delta_{fus}C_p^o = 0$), i.e., that the heat of fusion is independent of temperature. If this is so,

$$\Delta_{fus}G_T^o/2.303RT = \Delta_{fus}H^o(1 - T/T_{fus})/2.303RT \qquad (4.5)$$

Experimental heat of fusion data exist for a large number of compounds (Domalski and Hearing, 1996) or may be estimated by group contribution methods (Joback and Reid, 1987; Simamora et al., 1993). Eq. (4.5) may be rearranged to give

$$\Delta_{fus}G_T^o/2.303RT = \Delta_{fus}H^o(T_{fus} - T)/2.303RTT_{fus} \qquad (4.6)$$

which contains the expression $\Delta_{fus}H^o/T_{fus} = \Delta_{fus}S^o$. The entropy of fusion of rigid organic molecules such as simply substituted benzenes is approximately $56.5 \, J \, mol^{-1} \, K^{-1}$ (Yalkowsky, 1979). This is also true for a number of PCBs (Miller et al., 1984). With this approximation, the melting point correction is -0.25 for $T_{fus} = 50°C$ or -0.74 for $T_{fus} = 100°C$, at a reference temperature of $25°C$.

The melting point correction term is therefore a significant quantity which has been included (Bowman and Sans, 1983; Chiou et al., 1982; Isnard and Lambert, 1989; Mackay et al., 1980; Miller et al., 1984) or omitted (Banerjee, 1980; Chiou et al., 1977; Isnard and Lambert, 1989) in correlations. Whether or not inclusion is important to the final result depends on the quality of K_{ow} and the water solubility data used in the correlation.

2 HIGH PERFORMANCE LIQUID CHROMATOGRAPHY (RP-HPLC) CORRELATIONS

As mentioned in Chapter 3, this is a very widely used method for determining partition coefficients. It is not difficult to see why (Lambert, 1993). The entire procedure is speedy and no quantitative analysis is required. Only milligram quantities of sample are needed, and it need not be pure. The useful upper K_{ow} limit in RP-HPLC is higher than that of the shake-flask method and surface-active solutes pose no particular problem. Results are highly reproducible between laboratories.

Earlier surveys of RP-HPLC applied to K_{ow} measurement were provided by Braumann (1986) and Hafkenscheid and Tomlinson (1986). They also discussed the various factors influencing the quality of correlations. The type of survey by Braumann is presented here in updated form (Tables 4.4–4.7). In these Tables the parameters of the equation

$$\log K_{ow} = a \log k' + b \qquad (3.4)$$

(k' being the capacity factor) are given for a large number of reported correlations, together with statistical data and other information concerning the phases and compounds used. Following Braumann's example, Table 4.7 contains similar information for correlations using capacity factors extrapolated to zero co-solvent content.

The very large number of entries in Tables 4.4–4.7 simply underscores the popularity of this method; significant insights into the limitations and qualifications of its use are not so easily discerned. These have been examined often (Table 4.1); Lambert (1993) provides a succinct and up-to-date summary.

Ideally, the partitioning process in RP-HPLC should mimic as closely as possible that in the octanol–water system in order that Eq. (3.4) faithfully represent K_{ow} reference data. In practice, of course, it does no such thing. ODS supports might be visualized as presenting a surface 'brush' or 'fur' of alkyl chains to the mobile solvent (Braumann, 1986). The structure of this layer is not well characterized, and may depend upon the mobile phase composition. The ease with which solute molecules penetrate this layer would then change with the amount of organic modifier. Depending on the method of alkylating the weakly acidic silica gel surface, there may be residual unreacted silanol groups, more or less solvated by mobile phase water or methanol. These residual silanol groups will interact differently with hydrogen-bonding solutes than will the alkyl chains. Such heterogeneity in the stationary phase has long been suspected as contributing to the existence of outliers in correlation plots and differing responses of various classes of hydrogen-bonding solutes. Schantz (1986) offers a readable and detailed account of these matters. Silanol masking agents such as alkylamines have sometimes been added to the mobile phase,

Table 4.4 Reported linear RP-HPLC correlations for non-hydrogen bonding solutes; $\log K_{ow} = a \log k' + b$

n	a	b	r	s	F	Modifier	φ	pH	Stationary phase	$\log K_{ow}$ range min, max	Reference
Alkyl benzenes											
18	1.92	1.91	0.954	0.17	163	MeOH	70		ODS-Hypersil	2.13, 4.26	Smith (1981)(S)
18	2.08	2.19	0.975	0.13	307	MeOH	50		C_{22} Magnusil	2.13, 4.26	Smith (1981)(S)
18	2.25	2.23	0.971	0.14	265	MeOH	50		SAS-Hypersil	2.13, 4.26	Smith (1981)(S)
5	1.42	2.27	0.998	0.05		MeOH	70		C_{18}-SIL-X-5	2.13, 4.38	Harnisch et al. (1983)
10	2.08	2.63	0.977			MeOH	70		μBondapak RP-18		Koopmans and Rekker (1984)(B)
5	3.03	2.39	0.984			MeOH	70		Finesil RP-18		Jinno (1982)(B)
19	1.13	2.07	0.906			MeCN	75		Finesil RP-18		Jinno and Kawasaki (1983)(B)
15	4.66	1.92	0.990			MeCN	75				Whitehouse and Cook (1982)(B)
Polycyclic aromatic hydrocarbons											
26	3.20	3.24	0.977			MeOH	90		Partisil RP-18		Ruepert et al. (1985)(B)
10	3.92	1.34	0.985			MeCN	70		Develosil RP-18		Hanai and Hubert (1984)(B)
10	11.23	−3.31	0.901			THF	50		Develosil RP-18		Hanai and Hubert (1984)(B)

Non-hydrogen bonders

6	1.52	1.09	0.995	0.05		none	0		Corasil C-18*	3.64, 4.78	Lins et al. (1982)
6	2.23	2.30	0.962	0.14		MeOH	40		Corasil C-18	3.64, 4.78	Lins et al. (1982)
11	1.53	2.83	0.984	0.11		MeOH	10	2.2	Gly-CPG	2.13, 3.38	Miyake and Terada (1982)
13	2.37	1.76	0.990	0.13	521	MeOH	70		ShimPack CLC-ODS	2.11, 4.45	Sun et al. (1994)

Halogenated benzenes

13	3.61	1.43	0.988		MeCN	70		Develosil RP-18	Hanai and Hubert (1984)(B)

*coated with octanol
φ=volume percent modifier in mobile phase
Gly-CPG=glycerol-coated controlled pore glass beads
MeOH=methanol; MeCN=acetonitrile; THF=tetrahydrofuran

Table 4.5 Reported linear RP-HPLC correlations for H-bonding solutes; $\log K_{ow} = a \log k' + b$

n	a	b	r	s	F	Modifier	φ	pH	Stationary phase	$\log K_{ow}$ range min, max	Reference
Urinary aromatic compounds											
37	1.96	1.31	0.929			MeCN	40	2.0	Chromosorb RP-18	−0.01, 2.83	Hanai and Hubert (1982)\(B)
Aromatic acids											
12	1.26	1.61	0.985			MeOH	50	2.2	Hypersil RP-18		Hafkenscheid and Tomlinson (1983)\(B)
Benzoic acids											
16	2.01	2.47	0.987	0.09		MeOH	10	2.2	Gly-CPG	1.23, 3.14	Miyake and Terada (1982)
Aryloxoalkanoic acids											
17	1.66	0.95	0.992			MeOH	50	3.0	μBondapak RP-18	0.80, 3.46	Kuchař et al. (1985)\(B)
Phenoxyacetic acids											
7	2.31	3.40	0.941			MeOH	70	2.9	LiChrosorb RP-18	3.18, 3.66	Braumann et al. (1983)\(B)
Phenols											
4	1.68	0.897	0.997	0.09	368	MeOH	55	7.0	Alltech RP-18	1.63, 3.30	Haky and Young (184)\(S)
9	1.91	1.92	0.961			Acet			μBondapak RP-18	1.46, 2.59	Carlson et al. (1975)\(B)
26	2.08	2.50	0.988	0.14		MeOH	10	2.2	Gly-CPG	1.23, 5.12	Miyake and Terada (1982)
Hydroxybenzenes											
6	0.946	0.605	1.00	0.02	9633	none	0		Eurospher C-18*	0.80, 3.30	Ritter et al. (1994)
H-donors											
13	2.24	2.46	0.987	0.13	427	MeOH	70		ShimPack CLC-ODS	1.10, 3.72	Sun et al. (1994)(S)

H-acceptors											
9	2.35	2.65	0.993	0.12		MeOH	10	2.2	Gly-CPG	1.56, 4.21	Miyake and Terada (1982)
H-bonders											
39	1.43	0.146	0.974	0.23	705	MeOH	30	7.4	CapcellPak C$_{18}$	−0.77, 2.67	Yamagami and Takao (1992)(S)
Alcohols and ethers											
40	1.67	0.552	0.972	0.21	679	MeOH	50		Zorbax ODS	−0.32, 3.21	Funasaki et al. (1986)(S)
Benzamides											
10	1.90	0.790	0.987			MeCN	18		Nucleosil RP-18	0.79, 3.77	Verbiese-Gérard et al. (1981)(B)
Sulphonamides											
10	0.980	−3.03	0.937	0.27		none	0	4.0	C-18 Corasil	−1.22, 1.15	Henry et al. (1976)
10	1.37	0.39	0.862	0.39		none	0	4.0	Corasil II†	−1.22, 1.15	Henry et al. (1976)
10	1.38	0.250	0.863	0.39		none	0	4.0	Corasil II*	−1.22, 1.15	Henry et al. (1976)
Substituted benzenes											
42	1.72	1.34	0.932			MeOH	60		Nucleosil RP-18		Schoenmakers et al. (1981)(B)
41	1.90	1.34	0.893			MeCN	50		Nucleosil RP-18		Schoenmakers et al. (1981)(B)
25	2.21	1.31	0.817			THF	40		Nucleosil RP-18		Schoenmakers et al. (1981)(B)
Phenylureas											
11	2.31	2.62	0.925			MeOH	70		LiChrosorb RP-18	1.18, 4.31	Braumann et al (1983)(B)
6	1.32	1.29	0.950			MeOH	30		Micropak RP-18	1.91, 2.68	Rittich et al. (1984)(B)

Table 4.5 (Continued)

n	a	b	r	s	F	Modifier	φ	pH	Stationary phase	log K_{ow} range min, max	Reference
Benzophenones											
14	3.61	2.94	0.986			MeOH	70		μBondapak RP-18	3.19, 6.31	Kakoulidou and Rekker (1984)(B)
Pyridazinones											
8	1.61	1.79	0.956			MeOH	60		LiChrosorb RP-18		Braumann and Grimme (1981)(B)
N-Methyl carbamates											
9	4.53	−0.77	0.946			MeOH	25		Micropak RP-18	0.90, 2.03	Rittich et al. (1984)(B)
9	1.85	0.65	0.968			Diox	25		Micropak RP-18	0.90, 2.03	Rittich et al. (1984)(B)
Dimethyl benzylcarbamates											
19	1.64	0.962	0.998	0.05	3563	MeOH	50	7.4	CapcellPak C_{18}	0.77, 3.32	Yamagami and Takao (1993)(S)
N-Phenylsuccinimides											
11	2.42	2.61	0.984	0.11		MeOH	10	2.2	Gly-CPG	1.08, 2.80	Miyake and Terada (1982)
N-Phenylanthranilates											
9	2.16	2.87	0.983	0.13		MeOH	10	2.2	Gly-CPG	3.49, 5.57	Miyake and Terada (1982)
1-Arylpiperazines											
12	2.09	0.92	0.985			MeCN	70	5–7.7	μBondapak RP-18	1.10, 5.50	Fong et al. (1985)(B)
Anilines											
12	2.24	1.44	0.968			Acet			μBondapak RP-18	0.90, 2.69	Carlson et al. (1975)(B)

n											
Phenothiazines											
16	2.94	2.58	0.973	0.15		MeCN	70	7.5	μBondpak RP-18	1.95, 4.03	Barbato et al. (1982)
5-Nitroimidazoles											
22	2.76	−0.216	0.921	0.30	111	MeOH	40		μBondpak RP-18	−1.00, 2.03	Guerra et al. (1983)(S)
1-Naphthyloxyalkyl amines											
11	2.38	1.94	0.970	0.29	142	MeCN	40	7.4	RP-18*	1.01, 4.39	Unger et al. (1986)
2-Alkyl-5-amidobenzotriazoles											
8	2.09	2.09	0.994	0.12	524	MeOH	60		μBondpak RP-18	1.22, 5.85	Caliendo et al. (1991)

*coated with octanol
†coated with squalene
φ=volume per cent modifier in mobile phase
MeOH=methanol; MeCN=acetonitrile; Acet=acetone; THF=tetrahydrofuran; Diox=1,4-dioxane
For explanation of other symbols, see Table 4.2
(B) after the reference indicates statistical data are from Braumann (1986); (S), calculated by author (J. S.)

Table 4.6 Reported linear RP-HPLC correlations for miscellaneous compounds; log K_{ow} = a log k' + b

n	a	b	r	s	F	Modifier	φ	pH	Stationary phase	log K_{ow} range min, max	Reference
Barbiturates											
10	1.53	1.01	0.952			MeOH	50		Hypersil RP-18	0.69, 3.25	Hafkenscheid and Tomlinson (1983)(B)
24	3.58	0.52	0.945			MeOH	50		Partisil RP-18	0.65, 3.00	Wells et al. (1981)(B)
Hydantoins											
17	1.63	0.830	0.963			MeOH	50		Hypersil RP-18		Hafkenscheid and Tomlinson (1983)(B)
Pesticides											
21	2.32	1.88	0.976			EtOH	65		LiChrosorb RP-18	0.75, 6.34	Ellgehausen et al. (1981)(B)
Isoxazolyl penicillins											
8	1.27	1.83	0.965	0.10						1.80, 2.91	Thijssen (1981)
Basic Drugs											
29	1.91	0.970	0.974			MeOH	50	4 or 7	Hypersil RP-18	1.14, 5.90	Hafkenscheid and Tomlinson (1983)(B)
Various											
37	2.70	2.37	0.987			MeOH	75		Zorbax RP-18		McDuffie (1981)(B)
20	2.50	2.12	0.989			MeOH	70		LiChrosorb RP-18	0.93, 5.00	Könemann et al. (1979)(S)
68	1.65	0.948	0.966	0.26	925	MeOH	55	7.0	Alltech RP-18	0.17, 4.46	Haky and Young (1984)(S)
36	1.39	1.39				MeOH	50	4.5	LiChrosorb RP-18	1.83, 7.47	Hammers et al. (1982)(B)

n						modifier	φ	stationary phase		log P range		Reference
12	3.37	2.88	0.988	0.25		MeOH	85	Ultrasphere IPC$_{18}$				Rapaport and Eisenreich (1984)
21	1.09	1.15	0.982	0.09	513	TEA	1	Corasil C-18		1.32	2.69	McCall (1975)
21	1.08	1.28	0.927	0.18	116	MeCN	15	Corasil C-18		1.32	2.69	McCall (1975)
7	1.93	1.91	0.986	0.36		MeOH	70	C$_{18}$-SIL-X-5		−1.57	5.47	Harnisch et al. (1983)
10	1.04	1.51				none	0	Corasil I*	2	0.60	2.20	Miyake and Terada (1978)
82	2.26	2.58	0.984	0.19		MeOH	10	Gly-CPG	2.2	1.08	5.57	Miyake and Terada (1978)
11	1.02	0.775	0.994			none	0	ODS*	7.0	0.90	3.32	Caron and Shroot (1984)
33	1.03	0.797	0.987	0.13	1167	none	0	Corasil C$_{18}$*	7.0	−0.24	3.39	Unger et al. (1978)
5			0.998	0.03		MeOH	50	LiChrosorb RP-18	7.0	1.00	4.00	Hagen et al. (1989)

* coated with octanol

φ = volume per cent modifier in mobile phase

Gly-CPG = glycerol coated controlled pore glass beads

MeOH = methanol; EtOH = ethanol; TEA = tetraethyl amine; MeCN = acetonitrile

For explanation of other symbols, see Table 4.2

(B) after the reference indicates statistical data are from Braumann (1986); (S), calculated by author (J. S.)

Table 4.7 Reported linear RP-HPLC relationships (extrapolated capacity factor); log $K_{ow} = a \log k^w + b$

n	a	b	r	s	F	pH	Stationary phase	log K_{ow} range min, max	Reference
Alkyl benzenes									
5	0.870	0.050	0.998			W	C$_{18}$-SIL-X	2.13, 4.38	Harnisch et al. (1983)|(B)
14	0.937	0.265	0.999			W	PE HS-5-C$_{18}$	2.13, 3.90	Sherblom and Eganhouse (1988)
PCBs									
11	0.886	1.01	0.989			W	PE HS-5-C$_{18}$	4.00, 7.00	Sherblom and Eganhouse (1988)
Non-hydrogen bonders									
8	0.770	0.900	0.997	0.06	974	2.0	Finepak SIL C$_{18}$-T	2.13, 4.04	Miyake et al. (1986)
Polar benzenes									
12	0.930	−0.070				4.5	LiChrosorb RP-18		Hammers et al. (1982)|(B)
Substituted benzenes									
45	1.12	−0.360	0.974			W	LiChrosorb RP-18		Braumann et al. (1983)|(B)
25	1.02	−0.070	0.994			n.s.	n.s.		Braumann (1986)
43	1.07	−0.195	0.982	0.19	1123	7.4	LiChrosorb RP-18	−0.36, 3.42	El Tayar et al. (1985b)|(S)
6	1.06	−0.22	0.993	0.10	297	7.4	LiChrosorb RP-18	0.90, 2.84	El Tayar et al. (1985b)|(S)
Halogenated benzenes									
9	1.07	0.697	0.989	0.12	312	NF	octyl silane	1.85, 4.51	Vallat et al. (1992)|(S)
Phenols									
47	1.19	0.353	0.981	0.20	1122	1	LiChrosorb RP-18	0.55, 5.01	Butte et al. (1981)|(S)
13	1.18	−0.048	0.995	0.07	993	NF	octyl silane	1.46, 3.69	Vallat et al. (1992)|(S)
21	0.986	0.062	0.98	0.12	468				Xie et al. (1984)
Benzoic acids									
9	1.17	−0.355	0.975	0.11	133	NF	octyl silane	1.77, 3.02	Vallat et al. (1992)|(S)
H-Bonders									
61	1.01	−0.085	0.981	0.17	1483	7.4	CapcellPak C$_{18}$	−0.11, 3.22	Yamagami et al. (1994)|(S)

Group / n							Column		Reference
Amphiprotics									
25	0.730	1.00	0.979	0.17	531	2.0	Finepak SIL C_{18}-T	1.48, 4.36	Miyake et al. (1986)
Dimethyl benzylcarbamates									
18	1.08	−1.07	0.987	0.12	581	7.4	CapcellPak C_{18}	0.77, 3.32	Yamagami and Takao (1993)(S)
Azoxy cyanobenzenes									
25	1.16	−1.07	0.988	0.15		W	LiChrospher RP-18	−0.41, 3.17	Calvino et al. (1991)
Benzodiazepines									
9	1.12	0.14	0.968			9.0	Corasil-RP-18	1.18, 3.50	Hulshoff and Perrin (1976)(B)
OECD reference compounds									
46	0.90	0.08	0.983			W	C_{18}-SIL-X	−1.57, 5.47	Harnisch et al. (1983)(B)
Barbiturates									
18	1.06	−0.110	0.976			W	CO:PELL ODS		Wells et al. (1981)(B)
Drugs									
20	1.14	−0.998	0.937	0.56	129	W	LiChrosorb RP-18	0.58, 6.23	El Tayar et al. (1985a)
H-acceptors									
11	0.340	0.970	0.996	0.06	1266	2.0	Finepak SIL C_{18}-T	1.48, 3.59	Miyake et al. (1986)
Various									
36	0.910	0.280				4.5	LiChrosorb RP-18		Hammers et al. (1982)(B)
44	0.710	0.930	0.957	0.22	459	2.0	Finepak SIL C_{18}-T	1.48, 4.36	Miyake et al. (1986)
32	1.02	−0.060	0.991			W		0.55, 4.46	Hafkenscheid and Tomlinson (1981)
9	0.984	0.134	0.998			7.4	Chromegabond C_8	0.95, 4.47	Pomper et al. (1990)
34	0.852	0.374	0.986	0.12		7.4	ES C_8	0.80, 3.82	Minick et al. (1988)
14	1.02	0.021	0.997	0.12	1986	NF	Supelcosil LC-8-DB	0.89, 5.70	Jencke et al. (1990)(S)

Table 4.7 *(Continued)*

n	a	b	r	s	F	pH	Stationary phase	log K_{ow} range min, max	Reference
34	1.21	0.125	0.943	0.24	256	7.4	Phenyl μBondapak	2.92, 5.29	Minick et al. (1987)\(S)
34	1.05	0.082	0.968	0.18	484	7.4	ODS μBondapak	2.92, 5.29	Minick et al. (1987)\(S)
40	0.830	−0.060	0.982	0.23		NF	ODPVA copolymer		Vallat et al. (1992)
70	1.09	0.697	0.989	0.12	312	NF	octyl silane		Vallat et al. (1992)
17	1.18	0.870	0.988			NF	RP-18	1.46, 7.88	Jencke (1996)

W = aqueous solvent was unbuffered water
n.s. = not specified
NF = aqueous solvent at a pH to suppress ionization
(B) after the reference indicates statistical data are from Braumann (1986); (S), calculated by author (J. S.)
For explanation of other symbols, see Table 4.2

but they are not always effective and may introduce other unwanted interactions (Lambert, 1993).

One can make the ODS column resemble octanol by coating it with that substance (Taylor, 1990). The eluent then is octanol-saturated water or buffer. Care must be taken in the preparation of such columns, as the coating has a tendency not to stay put. With these provisos, however, good correlations can be obtained.

Much effort has been devoted to the use of extrapolated capacity factors, i.e., k' is measured over a range of co-solvent concentrations and extrapolated to 100% water, to obtain k_w:

$$\log K_{ow} = c \log k_w + d \tag{4.7}$$

with the expectation that the parameters in Eq. (4.7) would approach the values $c = 1$, $d = 0$. Braumann (1986) strongly promoted this approach, suggesting the functional equivalence of $\log K_{ow}$ and $\log k_w$. Hammers et al. (1982) and Harnisch et al. (1983) also recommend it, although Klein et al. (1988) do not and Kachař et al. (1983) are indifferent. Whatever advantages may be gained from the use of k_w, however, it significantly increases the total amount of work required and compromises one of the principal advantages of the method that is, the time factor (Lambert, 1993).

Accounting for the solute hydrogen-bonding effects in RP-HPLC correlations has been a puzzle. Kim and Kim (1995) added a hydrogen-bonding indicator variable to Eq. (3.4) and claimed an improved correlation. A better understanding of these effects has resulted from a linear solvation energy relationship approach to K_{ow} and k' (Abraham et al., 1994). In this approach, K_{ow} and k' are expressed as functions of various solute descriptors such as hydrogen-bond acidity and basicity, dipolarity/polarizability, etc. (see Chapter 5). They report that '. . . the blend of factors that influence $\log k'$ . . . is not the same as that which influences $\log K_{ow}$. In particular, solute hydrogen-bond acidity considerably influences $\log k'$, but has no effect on $\log K_{ow}$ . . .'. Hence, in using RP-HPLC for estimating $\log K_{ow}$, care must be taken to match the reference set of solutes to the 'unknowns'; the rationale (Abraham et al., 1994) confirms what had been found empirically in previous work (Rekker et al., 1991).

In summary, then, RP-HPLC may be used reliably for measuring $\log K_{ow}$ provided sufficient attention is paid to

— the choice of stationary phase
— the choice of mobile phase (modifier, pH)
— the choice of reference compounds for calibration
— the quality of K_{ow} reference data
— the possibility of using extrapolated capacity factors.

Assuming a best-case scenario, Braumann (1986) speculated that '. . . upper limiting partition coefficients of approximately 11–12 are within experimental reach.'. This has turned out to be overly optimistic; it may be questioned what meaning, if any, is to be given to a measured $\log K_{ow}$ value of 11–12.

3 THIN-LAYER CHROMATOGRAPHY (TLC) CORRELATIONS

The well-known advantages of the HPLC correlation method (Section 2) are equally, if not more so, applicable to the TLC correlation method (Tomlinson, 1975). Some reported correlations are presented in Table 4.8, and those involving extrapolated R_M values are given in Table 4.9.

For best results in TLC conditions, elution conditions should be chosen such that $0.2 < R_F < 0.8$ (Kaliszan, 1981; Tomlinson, 1975). This gives a range of R_M of less than 1.5 decades, whereas HPLC is capable of 3–4 decades. This limitation may be overcome by changing the composition of the mobile phase. Care must be taken that changes in the rank order of elution (as in HPLC) do not interfere in a series of compounds of different polarity. In TLC, 'streaking' of spots sometimes is unavoidable, due to overloading of solute for better visualization. The method for developing or otherwise rendering the spots visible must of course be chosen carefully and carried out correctly. Precise pH control during elution is more easily accomplished in HPLC than in TLC. As in the shake-flask method, the effect of the ionization of the solute needs special attention. Hydrogen-bonding effects in chromatographic correlations may differ according to the stationary and mobile phases concerned. These may introduce uncertainty in comparison of results from different systems.

4 MEASURED ACTIVITY COEFFICIENTS

At first sight this appears to be an attractive source of K_{ow} data, since the relationships involved are deceptively simple:

$$K_{ow} = f^w / f^{oct} \qquad (2.42)$$

using volume-based Henrian activity coefficients or

$$K_{ow} = \gamma^w V_w / \gamma^{oct} V_{oct} \qquad (4.8)$$

using mole fraction-based activity coefficients (the Vs being the molar volumes of the solvents). As discussed in Chapter 2, however, the measurement of infinite dilution activity coefficients is not a simple matter. This is particularly true for solids or other involatile compounds.

Table 4.8 Reported linear TLC and RP-TLC relationships; $\log K_{ow} = aR_M + b$

n	a	b	r	s	F	Modifier	φ	pH	Stationary phase	log K_{ow} range min, max	Reference
PAHs											
9	4.52	4.81	0.980	0.20		MeOH	95	W	Whatman KC-8	2.81, 7.10	Bruggeman et al. (1982)
Phenols											
7	0.980	−0.014	0.946	0.33		MeOH	50	"acid"	KC$_8$ plates	1.58, 4.23	Kossoy et al. (1992)
Cinnamic acids											
35	−1.71	1.75	0.984	0.19		(1)			(2)	−1.25, 0.70	Kuchař et al. (1974)
Phenylacetic acids											
9	2.05	2.39	0.986	0.11	250	Acet	50	3.4	Merck Keiselgel	1.75, 3.90	Kuchař et al. (1983)(S)
Penicillins											
8	0.961	1.71	0.996	0.05	779			3.0		−0.06, 1.18	Bird and Marshall (1971)
Acidic drugs											
11	1.14	−0.858	0.978	0.39		MeOH	50	"acid"	KC$_8$ plates	5.40, 10.4	Kossoy et al. (1992)
Drugs with OH and amino groups											
7	1.44	−2.09	0.980	0.22		MeOH	50	NF	KC$_8$ plates	4.27, 6.28	Kossoy et al. (1992)
Pyrido [1, 2-a] pyrimidinones											
13	1.45	0.235	0.977	0.10		MeOH	30	W	Merck Kieselgel	0.20, 2.38	Papp et al. (1982)
Morpholinylethyl esters of alkoxyphenyl carbamic acid											
6	−1.05	2.45	0.990	0.10		(3)			(4)	2.95, 3.57	Bachratá et al. (1979)
Quinolines											
6	1.08	−0.087	0.885	0.34		MeOH	50	"basic"	KC$_8$ plates	1.63, 3.02	Kossoy et al. (1992)

Table 4.8 *(Continued)*

n	a	b	r	s	F	Modifier	φ	pH	Stationary phase	log K_{ow} range min, max	Reference
N-Alkyl triphenylmethyl amines											
8	4.07	1.83	0.998	0.10	1557	Acet	70	W	(5)	0.70, 4.34	Boyce and Milborrow (1965)(S) Rekker (1984)
Various											
8	4.26	4.45	0.985	0.17		Acet	60	W	Merck RP-18	1.14, 3.66	Tsantili-Kakoulidou and Antoniadou-Vyza (1989)
15	3.39	3.15	0.960			Acet	70	W	Merck RP-18	1.16, 6.19	Renberg et al. (1985)

(1) Benzene was the mobile phase
(2) Filter paper impregnated with formamide and ammonium formate
(3) Heptane was the mobile phase
(4) Cellulose foil impregnated with formamide
(5) Silica gel G impregnated with liquid paraffin

$φ$=volume percent modifier in mobile phase
MeOH=methanol; Acet=acetone
W=aqueous solvent was unbuffered water
NF=aqueous solvent was buffer at pH to suppress ionization
(S) after a reference indicates statistical data were calculated by the author (J. S.)
For explanation of other symbols, see Table 4.2

Table 4.9 Reported linear TLC and RP-TLC relationships (extrapolated R_M); $\log K_{ow} = a R_M^\circ + b$

n	a	b	r	s	F	pH	Stationary phase	$\log K_{ow}$ range min, max	Reference
Substituted phenols									
28	0.858	0.319	0.966	0.22	382	2.0	Merck RP-18	0.55, 3.38	Butte et al. (1981)
Substituted naphthols									
16	1.44	0.928	0.938	0.30	102		(1) (2)	1.61, 4.56	Biagi et al. (1979)
Phenylacetic acids									
9	0.695	1.73	0.993	0.08	483	3.4	Merck Kieselgel (2)	1.75, 3.90	Kuchař et al. (1983)(S)
Prostaglandins (2)									
12	1.24	−0.340	0.954	0.21	101	7.0	Silica gel G	1.14, 3.35	Barbaro et al. (1985)
Prostaglandins (3)									
12	1.15	−0.284	0.891	0.32	39	7.0	Silica gel G	1.14, 3.35	Barbaro et al. (1985)
Cephalosporins									
6	1.40	0.639	0.995	0.06	364	7.4	(1)	0.39, 1.85	Biagi et al. (1969b)(S) Yamana et al. (1977)
Penicillins									
13	0.941	1.21	0.990	0.12	537	7.4	(1)	0.59, 3.10	Biagi et al. (1969a)(S) Yamana et al. (1977)
β-Blockers									
12	1.42	−0.890	0.990	0.18	1423	10.2	Merck RP-18 (3)	−0.65, 3.20	Barbato et al. (1991)

(1) Silica gel impregnated with silicone oil
(2) R_M extrapolated from acetone solutions
(3) R_M extrapolated from methanol solutions
For explanation of other symbols, see Table 4.2

5 CENTRIFUGAL PARTITION CHROMATOGRAPHY (CPC)

This highly mechanized and automatized method for measuring partition coefficients has a number of enthusiastic supporters (Chapter 3). The usual caveats applicable to the other correlation methods apply also to CPC. Care must be taken that operating conditions are optimized for the range of K_{ow} envisaged and requirements of proper hydrodynamic flow in the apparatus. At best, its accuracy approaches that of RP-HPLC.

6 OTHER METHODS

Electrometric titration is a useful method for determining simultaneously both K_{ow} and pK_a of acidic or basic compounds. It cannot be considered a routine procedure in most cases. Micellar electrokinetic chromatography has only recently been applied to K_{ow} measurement; its limitations have not yet been fully explored and consequently may be considered to be still in development.

7 K_{ow} MEASUREMENT OF VERY HYDROPHOBIC COMPOUNDS

The measurement of partition coefficients for compounds of log $K_{ow} \geqslant 5$ poses particular problems for the shake-flask method. Unwanted adsorption on surfaces, purity of materials, separation of the phases and analysis become critical. Thus for polycyclic aromatic hydrocarbons, polychloro- and polybromo biphenyls and chlorinated dibenzodioxins and -dibenzofurans, other methods are usually employed. Table 4.10 summarizes actual practice. Although the data are not always as concordant as one would wish, nevertheless it is possible to extract reliable values for many compounds (Sangster, 1993).

Some organic dyes and colorants are also very hydrophobic. Since they are considered to be non-biodegradable, it was of some concern to determine their toxicity and bioconcentration factors (BCF) (Anliker and Clarke, 1981; Anliker et al., 1981). The BCF can be more easily estimated from K_{ow}; however, the measurement of K_{ow} was not very successful (Anliker and Moser, 1987; Anliker et al., 1981; Hou et al., 1991; Yen et al., 1989). In large part this is because these dyes have exceptionally low solubility in water and octanol (Anliker and Moser, 1987; Baughman and Perenich, 1988; Baughman and Weber, 1991; Yen et al., 1989). As it turns out, the potential for bioconcentration of these compounds is much less than anticipated, due mainly to their extreme water insolubility (Moser and Anliker, 1991).

Table 4.10 Measurement of K_{ow} for very hydrophobic compounds

Type of compound	Method	References
Polycyclic aromatic hydrocarbons (PAHs)	SF	Alcorn et al. (1993); Karickhoff et al. (1979); Mallon and Harrison (1984); Means et al. (1980); Sanemasa et al. (1994) (isopiestic modification)
	SS	Brooke et al. (1990); De Bruijn et al. (1989)
	RP-HPLC	Brooke et al. (1986); D'Amboise and Hanai (1982); Kaune et al. (1995); Mallon and Harrison (1984); Rapaport and Eisenreich (1984); Sarna et al. (1984); Wang et al. (1986); Webster et al. (1985b)
	RP-TLC	De Voogt et al. (1990)
	sol'y (W)	Bruggeman et al. (1982); Mackay et al. (1980)
	sol'y (W, O)	Miller et al. (1985)
Polychloro-and polybromo-biphenyls (PCBs, PBBs)	SF	Chiou et al. (1977); Karickhoff et al. (1979); Paris et al. (1978); Platford (1982)
	SS	Bruggeman et al. (1982); De Bruijn et al. (1989)
	GC	Doucette and Andren (1987); Hawker and Connell (1988); Larsen et al. (1992); Li and Doucette (1993); Miller et al. (1984); Woodburn et al. (1984)
	RP-HPLC	Brooke et al. (1986); Brodsky and Ballschmiter (1988); De Kock and Lord (1987); McDuffie (1981); Opperhuizen et al. (1985); Rapaport and Eisenreich (1984); Risby et al. (1990); Shaw and Connell (1982); Sherblom and Eganhouse (1988); Webster et al. (1985a)
	GLC	Risby et al. (1990)
	sol'y (W)	Mackay et al. (1980)
Chlorinated dibenzodioxins and -dibenzofurans	SS	Marple et al. (1986); Sijm et al. (1989)
	GC	De Voe et al. (1981); Doucette and Andren (1987); Shiu et al. (1988)
	RP-HPLC	Burkhard and Kuehl (1986); Jackson et al. (1993); Sarna et al. (1984); Webster et al. (1985a)

GC = generator column
GLC = gas–liquid chromatography (two liquids)
RP-HPLC = reversed phase high performance chromatography
RP-TLC = reversed phase thin layer chromatography
SF = shake-flask
sol'y (W); sol'y (W, O) = water solubility correlation; solubility in water and octanol
SS = slow stirring

8 GENERAL REMARKS

Much effort has been expended in investigating the relative accuracies of different K_{ow} measurement methods. There have been a number of single laboratory, interlaboratory and ring tests of varying thoroughness (Brooke *et al.*, 1990; Eadsforth and Moser, 1983; Kishi and Hashimoto, 1989; Klein *et al.*, 1988) and critical evaluations of published data (Brooke *et al.*, 1990; Chessells *et al.*, 1991). The methods considered were those most commonly used: shake-flask, slow-stirring, RP-HPLC, RP-TLC and generator column. The most recent (Chessells *et al.*, 1991) included a number of very lipophilic compounds (to log $K_{ow} = 10$). The importance of the octanol–water partition coefficient has been recognized officially by agencies, such as the US Environmental Protection Agency (USEPA, 1985). The OECD set forth guidelines for the measurement of K_{ow} by the shake-flast (OECD, 1981) and RP-HPLC methods (OECD, 1989). Although all the activity mentioned in this paragraph originated in the assessment of environmental effects of organic chemicals, the first proposed application actually was in medicinal chemistry (Hansch and Fujita, 1964).

Despite the current popularity of octanol as a co-solvent in partitioning, other bulk solvents have been and are being used also. Rekker and Mannhold (1992) list some of these common (and not so common) solvents. Leahy *et al.* (1989) proposed propylene dipelargonate as a new standard solvent. This substance has a balance of proton donor/acceptor properties which makes it complementary to octanol for determining biologically relevant partition coefficients. The choice among these solvents is discussed by Leahy *et al.* (1989), Smith *et al.* (1975) and Taylor (1990).

Concurrently with (or in place of) octanol, various heterogeneous media are sometimes used to deduce partition coefficients for correlating absorption, membrane permeability or *in vivo* drug distribution. Liposomes (or phospholipid vesicles) are often chosen for this purpose: L-α-dimyristoylphosphatidylcholine (DMPC) is a representative source. Such liposome partition coefficients have been measured for, inter alia: anti-inflammatory drugs (Betageri *et al.*, 1996), nitroimidazoles (Betageri and Rogers, 1989), imidazoline derivatives (Choi and Rogers, 1990; Rogers and Choi, 1993), β-blockers (Betageri and Rogers, 1987 and 1988), phenothiazines (Ahmed *et al.*, 1981), PCBs and chlorinated aromatics (Gobas *et al.*, 1988) as well as various molecules (Alcorn *et al.*, 1991 and 1993; Anderson *et al.*, 1983; Austin *et al.*, 1995). Micelles from quaternary ammonium salts have sometimes substituted for octanol (Garrone *et al.*, 1992; González *et al.*, 1992). A complete survey of this topic is beyond the scope of this book.

A summary of the characteristics of the principal methods of partition coefficient measurement is given in Table 4.11.

Table 4.11 Characteristics of log K_{ow} measurement methods; (Danielsson and Zhang, 1996; Sangster, 1989)

Method	log K_{ow} range*	Applicability	Strengths/Advantages	Weaknesses/Disadvantages
Shake-flask (SF)	−2.5, 4.5	General neutral species	Direct method Reliable when precautions taken OECD reference method	Time-consuming Requires attention to many details of procedure Temperature control may be inefficient Careful analysis required
Slow-stirring or sit-flask	< 8	Suitable for very hydrophobic compounds	Direct method Especially reliable for very lipophilic compounds Temperature control easier than SF	Long equilibration times Careful analysis required
Generator column	2–8	General, except hydophilic species	Direct method No danger of emulsion Closed system Efficient temperature control	Column becomes stripped of solute. New column needed for each solute. Elaborate analytical equipment required
RP-HPLC	0–6	Neutral species only	OECD standard procedure Insensitive to impurities Temperature control efficient Fast, convenient Analysis not required	Indirect method Careful matching of reference and unknown samples required
RP-TLC	0–6	Neutral species only	Faster and more convenient than RP-HPLC Useful for rapid screening of related compounds	As for RP-HPLC. Less accurate than RP-HPLC, unless log K_{ow} range restricted
Centrifugal partition chromatography	−2.5, 4	General neutral species	Highly mechanized method No solid support Insensitive to impurities Analysis not required	Indirect method Temperature control inefficient Hardware intensive

*These ranges are approximate and may be extended in either direction in particular cases

REFERENCES

Abraham, M. H., Chadha, H. S. and Leo, A. J. (1994) *J. Chromatogr. A* **685**, 203–11.
Ahmed, M., Burton, J. S., Hadgraft, J. and Kellaway, I. W. (1981) *J. Membr. Biol.* **58**, 181–9.
Alcorn, C. J., Simpson, R. J., Leahy, D. E. and Peters, T. J. (1991) *Biochem. Pharmacol.* **42**, 2259–64.
Alcorn, C. J., Simpson, R. J., Leahy, D. E. and Peters, T. J. (1993) *Biochem. Pharmacol.* **45**, 1775–82.
Anderson, N. H., Davis, S. S., James, M. and Kojima, I. (1983) *J. Pharm. Sci.* **72**, 443–8.
Anliker, R. and Clarke, E. A. (1981) *Chemosphere* **9**, 595–609.
Anliker, R. and Moser, P. (1987) *Ecotoxicol. Environ. Saf.* **13**, 43–52.
Anliker, R., Clarke, E. A. and Moser, P. (1981) *Chemosphere* **10**, 263–74.
Austin, R. P., Davis, A. M. and Manners, C. N. (1995) *J. Pharm. Sci.* **84**, 1180–3.
Bachratá, M., Blešová, M., Grolichová, L., Bezáková, Ž. and Lukás, A. (1979) *J. Chromatogr.* **171**, 29–36.
Banerjee, S., Yalkowsky S. H. and Valvani, S. C. (1980) *Environ. Sci. Technol.* **14**, 1227–9.
Barbaro, A. M., Guerra, M. C., Pietrogrande, M. C. and Borea, P. A. (1985) *J. Chromatogr.* **347**, 209–28.
Barbaro, F., Caliendo, G., Cappello, B. and La Rotonda, M. I. (1991) *Pharmacochem. Libr.* **16**, 95–8.
Barbato, F., Recanatini, M., Silipo, C. and Vittoria, A. (1982) *Eur. J. Med. Chem.-Chim. Ther.* **17**, 229–34.
Baughman, G. L. and Perenich, T. A. (1988) *Environ. Toxicol. Chem.* **7**, 261–71.
Baughman, G. L. and Weber, E. J. (1991) *Dyes Pigm.* **16**, 291–71.
Berthod, A. (1995) *Chromatogr. Sci. Ser.* **68**, 167–97.
Berti, P., Cabani, S., Conti, G. and Mollica, V. (1986). *J. Chem. Soc.* Faraday Trans. I **82**, 2547–56.
Betageri, G. V. and Rogers, J. A. (1987) *Int. J. Pharm.* **36**, 165–73.
Betageri, G. V. and Rogers, J. A. (1988) *Int. J. Pharm.* **46**, 95–102.
Betageri, G. V. and Rogers, J. A. (1989) *Pharm. Res.* **10**, 913–17.
Betageri, G. V., Nayernama, A. and Habib, M. J. (1996) *Int. J. Pharm. Adv.* **1**, 310–19.
Biagi, G. L., Barbaro, A. M., Gamba, M. F. and Guerra, M. C. (1969a) *J. Chromatogr.* **41**, 371–9.
Biagi, G. L., Barbaro, A. M., Guerra, M. C. and Gamba, M. F. (1969b) *J. Chromatogr.* **44**, 195–8.
Biagi, G. L., Barbaro, A. M., Guerra, M. C., Solaini, G. C. and Borea, P. A. (1979) *J. Chromatogr.* **177**, 35–49.
Bird, A. E. and Marshall, A. C. (1971) *J. Chromatogr.* **63**, 313–19.
Bowman, B. T. and Sans, W. W. (1983) *J. Environ. Sci. Health B* **18**, 667–83.
Boyce, C. B. C. and Milborrow, B. V. (1965) *Nature (London)* **208**, 537–9.
Braumann, T. (1986) *J. Chromatogr.* **373**, 191–225.
Braumann, T. and Grimme, L. H. (1981) *J. Chromatogr.* **206**, 7–15.
Braumann, T., Weber, G. and Grimme, L. H. (1983) *J. Chromatogr.* **261**, 329–43.
Brodsky, J. and Ballschmiter, K. (1988) *Fresenius Z. Anal. Chem.* **331**, 295–301.
Brooke, D. N., Dobbs, A. J. and Williams, N. (1986) *Ecotoxicol. Environ. Saf.* **11**, 251–60.
Brooke, D. N., Nielsen, I., De Bruijn, J. and Hermens, J. (1990) *Chemosphere* **21**, 119–33.
Bruggeman, W. A., van der Steen, J. and Hutzinger, O. (1982) *J. Chromatogr.* **238**, 335–46.
Burkhard, L. P. and Kuehl, D. W. (1986) *Chemosphere* **15**, 163–7.

Butte, W., Fooken, C., Klussman, R. and Schuller, D. (1981) *J. Chromatogr.* **214**, 59–67.

Cabani, S., Conti, G., Mollica, V. and Bernazzani, L. (1991) *J. Chem. Soc.* Faraday Trans. *87*, 2433–42.

Caliendo, G., Novellino, E., Santagada, V., Silipo, C. and Vittoria, A. (1991) *Pharmacochem. Libr.* **16**, 401–4.

Calvino, R., Fruttero, R. and Gasco, A. (1991) *J. Chromatogr.* **547**, 167–73.

Carlson, R. M., Carlson, R. E. and Kopperman, H. L. (1975) *J. Chromatogr.* **107**, 219–23.

Caron, J. C. and Shroot, B. (1984) *J. Pharm. Sci.* **73**, 1703–6.

Chessells, M., Hawker, D. W. and Connell, D. W. (1991) *Chemosphere* **22**, 1175–90.

Chiou, C. T., Schmedding, D. W. and Manes, M. (1982) *Environ. Sci. Technol.* **16**, 4–10.

Chiou, C. T., Freed, V. H., Schmedding, D. W. and Kohnert, R. L. (1977) *Enviorn. Sci. Technol.* **11**, 475–8.

Choi, Y. W. and Rogers, J. A. (1990) *Pharm. Res.* **7**, 508–12.

Dallas, A. J. and Carr, P. W. (1992) *J. Chem. Soc. Perkin Trans,* 2, 2155–61.

D'Amboise, M. and Hanai, T. (1982) *J. Liq. Chromatogr.* **5**, 229–44.

Danielsson, L.-G. and Zhang, Y.-H. (1996) *Trends Anal. Chem.* **15**, 188–96.

Dearden, J. C. and Bresnen, G. M. (1988) *Quant. Struct.-Act. Relat.* **7**, 133–44.

De Bruijn, J., Busser, F., Seinen, W. and Hermens, J. (1989) *Environ. Toxicol. Chem.* **8**, 499–512.

De Kock, A. C. and Lord, D. A. (1987) *Chemosphere* **16**, 133–42.

De Voe, H., Miller, M. M. and Wasik, S. P. (1981) *J. Res. Nat. Bur. Stand. (U.S.)* **86**, 361–6.

De Voogt, P., van Zijl, G. A., Govers, H. and Brinkman, U. A. T. (1990) *J. Planar Chromatogr.-Mod. TLC* **3**, 24–33.

Domalski, E. S. and Hearing, E. D. (1996) *J. Phys. Chem. Ref. Data* **25**, 1–525.

Doucette, W. J. and Andren, A. W. (1987) *Environ. Sci. Technol.* **21**, 821–4.

Eadsforth, C. V. and Moser, P. (1983) *Chemosphere* **12**, 1459–75.

Ellgehausen, H., D'Hondt, C. and Fuerer, R. (1981) *Pestic. Sci.* **12**, 219–27.

El-Tayar, N., Marston, A., Bechalany, A., Hostettmann, K. and Testa, B. (1989) *J. Chromatogr.* **469**, 91–9.

El Tayar, N., van de Waterbeemd, H. and Testa, B. (1985a) *J. Chromatogr.* **320**, 305–12.

El Tayar, N. van de Waterbeemd, H. and Testa, B. (1985b) *Quant. Struct. Act. Relat.* **4**, 69–77.

El Tayar, N., Marston, A., Bechalany, A., Hostettman, K. and Testa, B. (1989) *J. Chromatogr.* **469**, 91–9.

Esser, H. O. and Moser, P. (1982) *Ecotoxicol. Enviorn. Saf.* **6**, 131–48.

Fong, M. H., Aarons, L., Urso, R. and Caccia, S. (1985) *J. Chromatogr.* **333**, 191–7.

Funasaki, N., Hada, S. and Neya, S. (1986) *J. Chromatogr.* **361**, 33–45.

Garrone, A., Marengo, E., Fornatto, E. and Gasco, A. (1992) *Quant. Struct.-Act. Relat.* **11**, 171–5.

Gobas, F. A. P. C., Lahittete, J. M., Garofalo, G., Shiu, W. Y. and Mackay, D. (1988) *J. Pharm. Sci.* **77**, 265–72.

González, V., Rogríguez-Delgado, M. A., Sánchez, M. J. and García-Montelongo, F. (1992) *Chromatographia* **34**, 627–35.

Guerra, M. C., Barbaro, A. M., Cantelli Forti, G., Biagi, G. L. and Borea, P. A. (1983) *J. Chromatogr.* **259**, 329–33.

Hafkenscheid, T. L. and Tomlinson, E. (1981) *J. Chromatogr.* **218**, 409–25.

Hafkenscheid, T. L. and Tomlinson, E. (1983) *Int. J. Pharm.* **16**, 225–39.

Hafkenscheid, T. L. and Tomlinson E. (1984) *J. Chromatogr.* **292**, 305–17.

Hafkenscheid, T. L. and Tomlinson, E. (1986) *Adv. Chromatogr.* **25**, 1–62.

Hagen, V., Hagen, A., Heer, S., Mitzner, R. and Niedrich, H. (1989) *Pharmazie* **44**, 20–3.

Haky, J. E. and Young, A. M. (1984) *J. Liq. Chromatogr.* **7**, 675–84.

Hammers, W. E., Meurs, G. J. and De Ligny, C. L. (1982) *J. Chromatogr.* **247**, 1–13.

Hanai, T. and Hubert, J. (1982) *J. Chromatogr.* **239**, 527–36.

Hanai, T. and Hubert, J. (1984) *J. Chromatogr.* **290**, 197–206.

Hansch, C., (gen. ed) (1990) *Comprehensive Medicinal Chemistry,* 6 vols., Pergamon Press, New York.

Hansch, C. and Fujita, T. (1964) *J. Am. Chem. Soc.* **86**, 1616–26.

Hansch, C., Quinlan, J. E. and Lawrence, G. L. (1968) *J. Org. Chem.* **33**, 347–50.

Harnisch, M., Möckel, H. J. and Schulze, G. (1983) *J. Chromatogr.* **282**, 315–32.

Hawker, D. W. and Connell, D. W. (1988) *Environ. Sci. Technol.* **22**, 382–7.

Henry, D., Block, J. H., Anderson, J. L. and Carlson, G. R. (1976) *J. Med. Chem.* **19**, 619–26.

Hou, M., Baughman, G. L. and Perenich, T. A. (1991) *Dyes Pigm.* **16**, 291–7.

Hulshoff, A. and Perrin, J. H. (1976) *J. Chromatogr.* **129**, 263–76.

Isnard, P. and Lambert, S. (1989) *Chemosphere* **18**, 1837–53.

Jackson, J. A., Diliberto, J. J. and Birnbaum, L. S. (1993) *Fundam. Appl. Toxicol.* **21**, 334–44.

Jencke, D. R. (1996) *J. Liq. Chromatogr. Rel. Technol.* **19**, 2227–45.

Jencke, D. R., Hayward, D. S. and Kenley, R. A. (1990) *J. Chromatogr. Sci.* **28**, 609–12.

Jinno, K. (1982) *Anal. Lett. A* **15**, 1533–7.

Jinno, K. and Kawasaki, K. (1983) *Chromatographia* **17**, 337–40.

Joback, K. G. and Reid, R. C. (1987) *Chem. Eng. Commun.* **57**, 233–243.

Kakoulidou, A. and Rekker, R. F. (1984) *J. Chromatogr.* **295**, 341–53.

Kaliszan, R. (1981) *J. Chromatogr.* **220**, 71–83.

Kaliszan, R. (1990) *Quant. Struct.-Act. Relat.* **9**, 83–7.

Karickhoff, S. W., Brown, D. S. and Scott, T. A. (1979) *Water Res.* **13**, 241–8.

Kaune, A., Knorrenschild, M. and Kettrup, A. (1995) *Fresenius J. Anal. Chem.,* **352**, 303–12.

Kim, K. H. and Kim, D. H. (1995) *Bioorg. Med. Chem.* **3**, 1389–96.

Kishi, H. and Hashimoto, Y. (1989) *Chemosphere* **18**, 1749–59.

Klein, W., Kördel, W., Weiß, M. and Poremski, H. J. (1988) *Chemosphere* **17**, 361–86.

Könemann, H., Zelle, R., Busser, F. and Hammers, W. E. (1979) *J. Chromatogr.* **178**, 559–65.

Koopmans, R. E. and Rekker, R. F. (1984) *J. Chromatogr.* **285**, 267–79.

Kossoy, A. D., Risley, D. S., Kleyle, R. M. and Nurok, D. (1992) *Anal. Chem.* **64**, 1345–9.

Kuchař, M., Brunova, B., Rejholec, V. and Rábek, V. (1974) *J. Chromatogr.* **92**, 381–9.

Kuchař, M., Kraus, E., Jelíková, M., Rejholec, V. and Miller, V. (1985) *J. Chromatogr.* **347**, 335–42.

Kuchař, M., Rejholec, V., Kraus, E., Miller, V. and Rábek, V. (1983) *J. Chromatogr.* **280**, 279–88.

Lambert, W. J. (1993) *J. Chromatogr. A* **656**, 469–84.

Larsen, B., Skejø-Andreasen, H. and Paya-Perez, A. (1992) *Fresenius Environ. Bull.* **1**, S13–S18.

Leahy, D. E. (1986) *J. Pharm. Sci.* **75**, 629–36.

Leahy, D. E., Taylor, P. J. and Wait, A. R. (1989) *Quant. Struct.-Act. Relat.* **8**, 17–31.

Leo, A. (1991) *Methods in Enzymology* **202**, 544–91.

Leo, A. J. (1995) *Chem. Pharm. Bull.* **43**, 512–13.

Li, A. and Doucette, W. J. (1993) *Environ. Toxicol. Chem.* **12**, 2031–5.

Lins, C. L. K., Block, J. H., Doerge, R. F. and Barnes, G. J. (1982) *J. Pharm. Sci.* **71**, 614–17.

Mackay, D., Bobra, A., Shiu, W. Y. and Yalkowsky, S. H. (1980) *Chemosphere* **9**, 701–11.

Mallon, B. J. and Harrison, F. L. (1984) *Bull. Environ. Contam. Toxicol.* **32**, 316–33.

Marple, L., Berridge, B. and Throop, L. (1986) *Environ. Sci. Technol.* **20**, 397–9.

McCall, J. M. (1975) *J. Med. Chem.* **18**, 549–52.

McDuffie, B. (1981) *Chemosphere* **10**, 73–83.

Means, J. C., Wood, S. G., Hassett, J. J. and Banwart, W. L. (1980) *Environ. Sci. Technol.* **14**, 1524–8.

Miller, M. M., Ghodbane, S., Wasik, S. P., Tewari, Y. B. and Martire, D. E. (1984) *J. Chem. Eng. Data* **29**, 184–90.

Miller, M. M., Wasik, S. P., Huang, G.-L., Shiu, W.-Y. and Mackay, D. (1985) *Environ. Sci. Technol.* **19**, 522–9.

Minick, D. J., Sabotka, J. J. and Brent, D. A. (1987) *J. Liq. Chromatogr.* **10**, 2565–89.

Minick, D. J., Frenz, J. H., Patrick, M. A. and Brent, D. A. (1988) *J. Med. Chem.* **31**, 1923–33.

Miyake, K. and Terada, H. (1978) *J. Chromatogr.* **157**, 386–90.

Miyake, K. and Terada, H. (1982) *J. Chro atogr.* **240**, 9–20.

Miyake, K., Mizuno, N. and Terada, H. (1986) *Chem. Pharm. Bull.* **34**, 4787–96.

Moser, P. and Anliker, R. (1991) *Proc. Int. Workshop Bioacc. Aquat. Syst.* 13–27.

OECD (1981) *Guidelines for Testing of Chemicals, No. 107*, Organization for Economic Co-operation and Development, Bureau of Information, Paris.

OECD (1989) *Guidelines for Testing of Chemicals, No. 117*, Organization for Economic Co-operation and Development, Bureau of Information, Paris.

Opperhuizen, A., van der Velde, E. W., Gobas, F. A. P. C., Liem, D. A. K., van der Steen, J. M. D. and Hutzinger, O. (1985) *Chemosphere* **14**, 1871–96.

Papp, O., Valko, K., Szasz, G., Hermecz, I., Vamos, J., Hanko, K. and Ignath-Halasz, Z. (1982) *J. Chromatogr.* **252**, 67–75.

Paris, D. F., Steen, W. C. and Baughman, G. L. (1978) *Chemosphere* **7**, 319–25.

Platford, R. F. (1982) *J. Great Lakes Res.* **8**, 307–9.

Pomper, M. G., Van Brocklin, H., Thieme, A. M. *et al.* (1990) *J. Med. Chem.* **33**, 3143–55.

Rapaport, R. A. and Eisenreich, S. J. (1984) *Environ. Sci. Technol.* **18**, 163–70.

Rekker, R. F. (1984) *J. Chromatogr.* **300**, 109–25.

Rekker, R. F. and Mannhold, R. (1992) *Calculating Drug Lipophilicity*, VCH Verlagsgesellschaft, Weinheim.

Rekker, R. F., de Vries, G. and Koopman, R. E. (1991) *Sci. Total Environ.* **109–110**, 179–95.

Renberg, L. O., Sundström, S. G. and Rosén-Olofsson, A.-C. (1985) *Toxicol. Environ. Chem.* **10**, 333–49.

Risby, T. H., Hsu, T.-B., Sehnert, S. S. and Bhan, P. (1990) *Environ. Sci. Technol.* **24**, 1680–7.

Ritter, S., Hauthal, W. H. and Maurer, G. (1994) *J. Chem. Eng. Data* **39**, 414–17.

Rittich, B., Štohandl, J., Žaludová, R. and Dubský, H. (1984) *J. Chromatogr.* **294**, 344–8.

Rogers, J. A. and Choi, Y. W. (1993) *Pharm. Res* **10**, 913–17.

Ruepert, C., Grinwis, A. and Govers, H. (1985) *Chemosphere* **14**, 279–91.

Sanemasa, I., Wu, J.-S. and Deguchi, T. (1994) *Anal. Sci.* **10**, 655–7.

Sangster, J. (1989) *J. Phys. Chem. Ref. Data* **18**, 1111–229.

Sangster, J. (1993) *LOGKOW—a Databank of Evaluated Octanol-water Partition Coefficients*, Sangster Research Laboratories, Montreal.

Sarna, L. P., Hodge, P. E. and Webster, G. R. B. (1984) *Chemosphere* **13**, 975–83.

Schantz, M. M. (1986) *Ph.D. Thesis*, Georgetown University.

Schantz, M. M. and Martire, D. E. (1987) *J. Chromatogr.* **391**, 35–51.

Schoenmakers, P. J., Billiet, H. A. H. and De Galan, L. (1981) *J. Chromatogr.* **218**, 261–84.

Shaw, G. R. and Connell, D. W. (1982) *Aust. J. Mar. Freshw. Res.* **33**, 1057–70.

Sherblom, P. M. and Eganhouse, R. P. (1988) *J. Chromatogr.* **454**, 37–50.

Shiu, W.-Y., Doucette, W., Gobas, F. A. P. C., Andren, A. and Mackay, D. (1988) *Environ. Sci. Technol.* **22**, 651–8.

Sijm, D. T. H. M., Wever, H., De Vries, P. J. and Opperhuizen, A. (1989) *Chemosphere* **19**, 263–6.

Simamora, P., Miller, A. H. and Yalkowsky, S. H. (1993) *J. Chem. Inf. Computer Sci.* **33**, 347–440.

Smith, R. M. (1981) *J. Chromatogr.* **209**, 1–6.

Smith, R. N., Hansch, C. and Ames, M. M. (1975) *J. Pharm. Sci.* **64**, 599–606.

Sun, X., Xin, M. and Zhao, J. (1994) *J. Liq. Chromatogr.* **17**, 1183–94.

Taylor, P. J. (1990) In Hansch (1990), Vol. 4, pp. 241–94.

Terada, H. (1986) *Quant. Struct.-Act. Relat.* **5**, 81–8.

Tewari, Y. B., Miller, M. M., Wasik, S. P. and Martire, D. E. (1982) *J. Chem. Eng. Data* **27**, 451–4.

Thijssen, H. H. W. (1981) *Eur. J. Med. Chem.-Chim. Ther.* **16**, 449–52.

Tomlinson, E. (1975) *J. Chromatogr.* **113**, 1–45.

Tsantili-Kakoulidou, A. and Antoniadou-Vyza, A. (1989) *Prog. Clin. Biol. Res.* **291**, 71–4.

Unger, S. H., Cook, J. R. and Hollenberg, J. S. (1978) *J. Pharm. Sci.* **67**, 1364–7.

Unger, S. H., Cheung, P. S., Chiang, G. H. and Cook, J. R. (1986) in W. H. Dunn, J. H. Block and R. S. Pearlman (eds) *Partition Coefficient Determination and Estimation*, Pergamon Press, New York (1986), pp. 69–81.

USEPA (1985) U.S. Environmental Protection Agency, Federal Register, Vol. 50, No. 188.

Vallat, P., Fan, W., El Tayar, N., Carrupt, P.-A. and Testa, B. (1992) *J. Liq. Chromatogr.* **15**, 2133–5.

Valvani, S. C., Yalkowsky, S. H. and Rosenman, T. J. (1981) *J. Pharm. Sci.* **70**, 502–7.

Verbiese-Gérard, N., Hanocq, M., van Damme, M. and Molle, L. (1981) *Int. J. Pharm.* **9**, 295–303.

Wang, L., Wang, X., Xu, O. and Tian, L. (1986) *Huanjing Kexue Xuebao* **6**, 491–7.

Wasik, S. P., Tewari, Y. B., Miller, M. M. and Martire, D. E. (1981) *Octanol/water Partition Coefficients and Aqueous Solubilities of Organic Compounds*. NBSIRR 81-2406, U.S. Department of Commerce, Washington.

Webster, G. R. B., Sarna, L. P. and Muir, D. C. G. (1985a) in *Chlorinated Dioxins and Dibenzofurans in the Total Environment*, ed. L. H. Keith, C. Rappe and G. Choudhary, Butterworths, London, pp. 79–87.

Webster, G. R. B., Friesen, K. J., Sarna, L. P. and Muir, D. C. G. (1985b) *Chemosphere* **14**, 609–22.

Wells, M. J. M., Clark, C. R. and Patterson, R. M. (1981) *J. Chromatogr. Sci.* **19**, 573–82.

Whitehouse, B. G. and Cook, R. C. (1982) *Chemosphere* **11**, 689–99.

Woodburn, K. B., Doucette, W. J. and Andren, A. W. (1984) *Environ. Sci. Technol.* **18**, 457–9.

Xie, T. M., Hulthe, B. and Folestad, S. (1984) *Chemosphere* **13**, 445–59.

Yalkowsky, S. H. (1979) *Ind. Eng. Chem. Fund.* **18**, 108–11.

Yamagami, C. and Takao, N. (1992) *Chem. Pharm. Bull.* **40**, 925–9.

Yamagami, C. and Takao, N. (1993) *Chem. Pharm. Bull.* **41**, 694–8.

Yamagami, C., Yokota, M. and Takao, N. (1994) *J. Chromatogr. A* **662**, 49–60.

Yamana, T., Tsuji, A., Miyamoto, E. and Kubo, O. (1977) *J. Pharm. Sci.* **66**, 747–9.

Yen, C.-P. C., Perenich, T. A. and Baughman, G. L. (1989) *Environ. Toxicol. Chem.* **8**, 981–6.

CHAPTER 5

Methods of Calculating Partition Coefficients

Although experimental $\log K_{ow}$ data exist for more than 18 000 organic chemicals (Sangster, 1993), this number is miniscule compared to the total number of compounds for which data are desirable. Besides, not all reported data are equally reliable. Hence there has been continuing interest in creating methods of calculating $\log K_{ow}$ from structure.

Some (though not all) of these methods take as their starting point the atoms or groups of atoms in the molecule, and $\log K_{ow}$ is derived from an additive–constitutive procedure for summing contributions of various atoms, groups or other structural information. Such an approach is indeed a common one for estimating physico-chemical properties of organic compounds (Barton, 1983; Lyman *et al.*, 1982; Reid *et al.*, 1987).

In this chapter, a number of calculational methods will be described, with a minimum of discussion and comparison. Their predictive performance will be examined in Chapter 6. Various methods have been reviewed by Hansch and Leo (1995), Leo (1990, 1993) and Taylor (1990). Some methods mentioned here exist in generally available computerized form. These are part of software packages which have many data handling features and other related capabilities. In the present context, only information relating to K_{ow} prediction is mentioned. Readers interested in further details should refer to the cited literature or contact the distributors at the addresses given.

1 FRAGMENT OR GROUP CONTRIBUTION METHODS

1.1 THE SUBSTITUENT CONSTANT π

It was noticed at an early date (Fujita *et al.*, 1964) that the difference between the partition coefficient of benzene (R-H) and simply substituted benzenes (R-X) was a characteristic quantity:

$$\pi_X = \log K_{ow}(\text{R-X}) - \log K_{ow}(\text{R-H}) \tag{5.1}$$

where X is the substituent. This suggested a more general method of predicting $\log K_{ow}$ for other molecules, e.g.,

$$\log K_{ow}(\text{Y-R-X}) = \log K_{ow}(\text{H-R-H}) + \pi_X + \pi_Y \tag{5.2}$$

This method has been described many times (Hansch and Leo, 1979, 1995; Leo, 1990, 1991, 1993; Leo *et al.*, 1971) and an extensive list of π-values has been accumulated (Hansch and Leo, 1995; Leo, 1990, 1991). This is an empirical method with rather obvious limitations, since in general the π-value of a substituent depends upon the number and identity of other substituents present.

1.2 THE METHOD OF HANSCH AND LEO

This has been outlined by the originators several times (Hansch and Leo, 1979, 1995; Leo, 1990, 1991, 1993). In the fragment method, the molecule is regarded as being constituted of a number of chemically recognizable and common atoms or groups of atoms. With the use of a large database of reliable experimental $\log K_{ow}$ data, contributions of fragments to the total $\log K_{ow}$ of molecules were established.

The strategy adopted by Hansch and Leo has been called a 'constructionist' approach; it apparently evolved from the π-concept described above. That is, the basic fragments were derived from a small set of the simplest possible molecules: for example, the partition coefficients of hydrogen, methane and ethane:

$$f(\text{H}) = \log K_{ow}(\text{H}_2)/2 \tag{5.3}$$

$$f(\text{CH}_3) = \log K_{ow}(\text{CH}_4) - f(\text{H}) \tag{5.4}$$

$$f(\text{CH}_3) = \log K_{ow}(\text{C}_2\text{H}_6)/2 \tag{5.5}$$

where f is a fragmental contribution to the total $\log K_{ow}$. The two values of $f(\text{CH}_3)$ from Eqs. (5.4) and (5.5) were the same within experimental error. The fragmental contributions of other atoms or groups of atoms followed in the usual manner (Cl, OH, CF_3, COOH, NO_2, etc.).

Two important observations occurred early on. The first, as mentioned before, was the finding that the value of a fragmental contribution—$f(\text{OH})$ say—depended to some extent on the nature of the chemical environment of the hydroxyl group. By 'chemical environment' is meant whether the attached carbon atom was aliphatic or aromatic, what the other substituents in the molecule were and how far away, the presence of carbon–carbon unsaturation, etc. (Fujita, 1983; Leo, 1983). Appropriate correction factors were derived for these effects. Second, the contribution of an alkyl side chain depended on both its length and the degree of branching, e.g., the contributions for n-butyl, isobutyl, sec-butyl and tert-butyl were all distinguishable.

The calculated log K_{ow} of a molecule thus could be represented by

$$\log K_{ow} = \Sigma a_n f_n + \Sigma b_m F_m \tag{5.6}$$

where a is the number of occurrences of fragment of type n and b is the number of occurrences of correction factor F of type m. A detailed discussion of the rationale of accounting for structural effects in this method is beyond the scope of this book; among them are electronic delocalization, steric, ortho-meta-para and intramolecular effects. The method is updated as new fragment constants and correction factors are determined (Calvino *et al.*, 1992).

The complete Hansch and Leo algorithm is called ClogP and is very complicated. It is beyond convenient manual execution; Hansch and Leo (1979) list 200 fragments and 25 correction factors, some generic. ('Generic' here simply means that the factor may be applied in many particular instances.) A simplified ClogP method is given in Lyman *et al.* (1982). The ClogP computer program is available from

> Biobyte Corporation
> Suite 204
> 201 West 4th Street
> Claremont, California 91711
> USA

1.3 THE METHOD OF REKKER ET AL.

The fragment method was actually first proposed and worked out by Rekker and colleagues. Theirs has been called a 'reductionist' approach, since the first step was to deduce a limited number of single- or few-atom contributions by direct analysis of reliable log K_{ow} data of molecules. As in Hansch and Leo's method, it was found that various structural factors also had to be taken into consideration, such as aryl–alkyl differences, proximity effects, chain length, etc. (Nys and Rekker, 1973, 1974). The final formula is again that of Eq. (5.6). In contrast to the Hansch and Leo method, the structural factors are attributed to a fundamental property of the water in the first solvation shell of the molecule; these factors are 'quantized', i.e., they are assigned an integral number times the so-called 'magic constant'.

The method is set forth in Rekker (1977) and in Rekker and de Kort (1979). Recently, it has been revised and updated (Rekker and Mannhold, 1992). Apart from this, it is apparently not updated on a more regular basis. At present (Rekker and Mannhold, 1992) there are 136 fragments and 10 structural correction factors (some generic). For consistent results, it requires a great deal of experience and skill for manual use. It is available in a computerized version called PrologP (Takeuchi *et al.*, 1990) from

CompuDrug Chemistry Limited
62 P.O.B., 405
H-1395 Budapest
Hungary

AR Software Corporation
Suite 1110
8201 Corporate Drive
Landover, Maryland 20785
USA

PrologP calculates neutral compounds only; zwitterionic molecules may be treated by a companion program, PrologD.

1.4 THE ACD METHOD

This algorithm, formulated recently, is generally similar to those already described. It uses 532 group contributions, 21 carbon atom type contributions and 2206 intramolecular correction factors. If the molecule contains groups or potential intramolecular interactions which are not included in the principal list, these are estimated by a number of secondary algorithms. This method has not been described in the scientific literature; it is updated as new information becomes known (A. Petrauskas, private communication, 1996). The method is known as ACD/LogP and is available from

Advanced Chemistry Development, Inc.
Suite 1501
141 Adelaide St W.
Toronto, Ontario
Canada M5H 3L5

1.5 THE METHOD OF MEYLAN AND HOWARD

In an approach quite similar to those above, Meylan and Howard (1995) devised 'atom/fragment contributions' (AFC), based upon a large up-to-date collection of reliable log K_{ow} data. The method is referred to as 'reductionistic' by the authors themselves; it uses 130 atom/fragment substructures and 235 correction factors. The general equation is

$$\log K_{ow} = \Sigma f_i n_i + \Sigma c_j n_j + 0.229 \tag{5.7}$$

similar to Eq. (5.6): atom/fragment contributions are indicated by f and correction factors by c. The method is updated as new contributions and correction factors are uncovered (W. Meylan, private communication, 1996). The AFC method is available in computerized form (KOWWIN) from

> Syracuse Research Corporation
> Environmental Science Center
> Syracuse, New York 13210
> USA

1.6 THE METHOD OF SUZUKI AND KUDO

These authors (Suzuki and Kudo, 1990) decided to create a group contribution method without correction terms:

$$\log K_{ow} = \Sigma n_i G_i \tag{5.8}$$

As might be expected, the groups were defined to take into account nearest bonded neighbours, as necessary. They tabulated 415 basic groups, 9 atom groups and 70 'extended groups'. These last are fragments containing one or more aliphatic or aromatic rings. A revised version was in preparation at time of writing.

2 ATOMISTIC METHODS

2.1 THE METHOD OF BROTO ET AL.

Rather than considering fragments, atomistic methods are based on contributions of single atoms. Of course, these single-atom contributions depend upon the local environment within the molecule. Broto *et al.* (1984) deduced a set of 222 contributions, including a conjugation term but none for internal hydrogen bonds. The algorithm needs computerized application, though none is generally available.

2.2 THE METHOD OF MORIGUCHI ET AL.

This may be considered an atomistic model based on the principle of counting the number of specified atoms (Moriguchi *et al.*, 1992). There are also contributions for proximity effects, unsaturation, etc. There are 13 parameters in all and the algorithm is simple enough for manual execution.

2.3 THE METHOD OF GHOSE AND CRIPPEN

Ghose and Crippen (1986) and Ghose *et al.* (1988), in a manner similar to Broto, deduced contributions for 110 atom types. This was extended later to 120 types (Viswanadhan *et al.*, 1989). The algorithm is beyond convenient manual execution and is available in computerized form as part of the PrologP program (see 'fragment method of Rekker *et al.*' above).

2.4 THE METHOD OF KLOPMAN ET AL.

This group began with a solvation model using the charge densities on each atom. A small number of parameters were deduced (Klopman and Iroff, 1981). A few group parameters were also introduced (Klopman *et al.*, 1985). The method was revised with the use of a much larger training set; 10 atomic parameters and 76 'star-centred fragments' were included (Klopman and Wang, 1991). The method was further upgraded recently (Klopman *et al.*, 1994) with an even larger training set. It is called by the authors an 'extended group contribution approach'. The parameter list contains atoms, functional groups and correction factors (overall total, 98). At this stage, the method — superficially at least — resembles the fragment methods already considered. A computerized version, called KlogP, is available from

> Multicase Inc.
> P.O. Box 22517
> Beachwood, Ohio 44122
> USA

and is updated as new experimental and theoretical findings come to light (G. Klopman, private communication, 1996).

3 'WHOLE MOLECULE' APPROACHES

In Chapter 2 it was seen that log K_{ow} is directly related to the Gibbs energy of transfer between the phases and also to the Henrian activity coefficients, Eqs. (2.42) and (2.61). It was also seen that by far the most significant contribution to the Gibbs energy of transfer is the Henrian activity coefficient in water, i.e., how the solute interacts with water in solution. This interaction is sometimes characterized by the Gibbs energy of solvation (gas→solution) (Pearlman, 1986). When a solute molecule enters a water environment, it takes up space and there are consequent solute–water and water–water interactions which to some extent are dependent upon the chemical nature of the solute.

A great deal of effort has been devoted to relating this Gibbs energy of interaction to solute molecular properties. Two of the simplest properties are molecular surface area and volume (Pearlman, 1986). Volume as a correlation parameter has been used principally in combination with other properties (see below). Surface area has been investigated as a single parameter to some degree.

3.1 SURFACE AREA

The first question which poses itself here is: which area do we want? Pearlman (1986) distinguishes three types: (1) van der Waals surface, (2) solvent-accessible surface area and (3) contact surface. The first is essentially the area of a space-filling solute molecule using Bondi radii. The second is the locus of the centre of a solvent molecule (approximated as a sphere) as it is rolled over the van der Waals surface. The third is defined by the same rolling procedure, but the locus here is that of the point, on the perimeter of the solvent sphere, which is closest to the van der Waals surface.

Since solutes will in general consist of polar and non-polar parts, the van der Waals surface is sometimes calculated for these types separately (Takács-Novák *et al.*, 1992). The surface area of hydrocarbon moieties is often called 'isotropic surface area'. The concept was further refined with the introduction of the 'supermolecule', i.e., the bare solute molecule together with its waters of hydration (Koehler *et al.*, 1988). A representative list of correlations of log K_{ow} with surface area is given in Table 5.1.

3.2 CONNECTIVITY (TOPOLOGY, GRAPH THEORY)

Topology concerns properties and spatial relations unaffected by continuous change of shape or size. In relation to molecules, *connectivity* deals with which atoms are connected to which other atoms. Since this is one unambiguous feature of well-defined organic molecules, molecular connectivity indices may be deduced directly from molecular structure (Kier, 1980; Kier and Hall, 1976). The indices of Kier and Hall are widely used. Others may be cited: Wiener

Table 5.1 Log K_{ow} correlations with surface area

Reference	Types of compounds (number of compounds)
Camilleri *et al.* (1988)	Various (217)
Chastrette *et al.* (1990)	Various (102)
Doucette and Andren (1987)	PCBs (17), PBBs (6), PCDDs (4), PCDFs (3)
Doucette and Andren (1988)	PCBs (26), PBBs (6), PCDDs (4), PCDFs (3), halobenzenes (13), alkyl benzenes (10)
Dunn *et al.* (1987)	Various (50)
Iwase *et al.* (1985)	Various (138)
Koehler *et al.* (1988)	Various (72)
Takács-Novák *et al.* (1992)	Imidazo quinolones (10)
Yalkowsky and Valvani (1976)	Hydrocarbons (14)

PCBs=polychlorinated biphenyls
PBBs=Polybrominated biphenyls
PCDDs=polychlorinated dibenzodioxins
PCDFs=polychlorinated dibenzofurans

indices (Lukovits, 1990; Pyka, 1995), characteristic root index (Saçan and İnel, 1995) and composites (Finizio *et al.*, 1995; Niemi *et al.*, 1992). Examples of such applications are given in Table 5.2.

3.3 MOLECULAR PROPERTIES FROM MOLECULAR ORBITAL CALCULATIONS

It is the fond hope of a great many current investigators that solvation Gibbs energies can be adequately estimated by *a priori* calculation of properties such as

> ionization potentials
> dipole moments
> electrostatic potentials
> charge densities
> charge-transfer energies
> highest occupied molecular orbital energies (HOMO)
> lowest unoccupied molecular orbital energies (LUMO)

This is only a partial list. These quantities are often used in combination with more mundane properties (surface area, volume). Much of the activity in this domain is due to the fact that computerized algorithms are generally available, along with computers of sufficiently high speed and capacity.

Table 5.2 Correlation of $\log K_{ow}$ with connectivity indices

Reference	Types of compounds (number of compounds)
Doucette and Andren (1988)	PCBs (26), PBBs (6), PCDDs, (4), PCDFs (3), halobenzenes (13), alkyl benzenes (10)
Finizio *et al.* (1994)	Chlorinated benzenes (17), chlorinated anilines (14)
Finizio *et al.* (1995)	Substituted benzenes (20), substituted anilines (22) agrochemicals (83)
Güsten *et al.* (1991)	PAHs (21), alkyl PAHs (22)
Güsten and Sabljić (1993)	Heterocyclics, PAHs (31)
Lukovits (1990)	Hydrocarbons (70)
Niemi *et al.* (1992)	Various (4067)
Parker (1978)	Hydroxylureas (8)
Patil (1991)	PCBs (140)
Pyka (1995)	Substituted phenols (14)
Saçan and İnel (1995)	PCBs (58)
Sabljić *et al.* (1993)	PCBs (20), chlorinated benzenes (13)
Szász *et al.* (1983)	2- and 3-ring nitrogen bridge compounds (56)

PCBs=polychlorinated biphenyls
PBBs=polybrominated biphenyls
PCDDs=polychlorinated dibenzodioxins
PCDFs=polycholorinated dibenzofurans
PAHs=polynuclear aromatic hydrocarbons

One of the earlier efforts in this direction was the 'solvent-dependent conformational analysis procedure' (SCAP) of Hopfinger and Battershell (1976). A much more powerful method, AM-1, was used by Bodor and co-workers (Bodor and Huang, 1991, 1992; Bodor *et al.*, 1989). AM-1 was used recently by other investigators with restricted classes of compounds (Makovskaya *et al.*, 1995; Schüürmann, 1995a,b). Other computational models, such as SCF, MNDO and PM3 have been used with larger sets of compounds, not necessarily restricted to a single class (Alkorta and Villar, 1992; Brinck *et al.*, 1993; Kantola *et al.*, 1991; Kasai *et al.*, 1988; Katagi *et al.*, 1995; Sasaki *et al.*, 1991).

In *comparative molecular field analysis* the steric, electrostatic and hydrophobic potential energy fields of molecules are calculated at various lattice points surrounding the molecule with the use of a CH_3, H^+, or H_2O probe group. Partition coefficients of PCBs (Waller, 1994) and heteroatom aromatic compounds (Kim, 1995; Kim and Kim, 1995) were deduced.

4 OTHER METHODS

4.1 UNIFAC ACTIVITY COEFFICIENTS

The acronym UNIFAC signifies UNIQUAC Functional Group Activity Coefficients model. It was developed by Fredenslund *et al.* (1975) and is based on a two-fluid model for describing a binary non-electrolyte mixture. Activity coefficients are calculated as a sum of two parts: a combinatorial portion (accounting for differences in size and shape of molecules) and a residual portion (representing interaction between functional groups). They are expressed in terms of group surface area and volume parameters which are found from experimental vapour–liquid equilibrium data. The formalism is easily adapted to multicomponent mixtures.

On the face of it, this method looks attractive; the mutual saturation of the solvents can easily be accounted for. In practice, its predictive accuracy is not particularly impressive (Arbuckle, 1983; Banerjee and Howard, 1988; Campbell and Luthy, 1985; Dallos *et al.*, 1993). This may be due to the fact that the UNIFAC model does not consider intramolecular interactions (Campbell and Luthy, 1985).

4.2 SCALED PARTICLE THEORY (SPT)

The SPT model of solute solvation visualizes solution as a two-step process: first, a cavity is formed in the solvent to accommodate the solute molecule; once inserted, the solute interacts with the solvent with the production of dispersion, inductive and dipole–dipole effects (Pierotti, 1976). Since the theory does not take into account specific interactions such as hydrogen bonding and charge transfer, the number of solutes which can be treated this way is restricted (Watarai *et al.*, 1982). 'Hydrophobic interactions' were added for hydrocarbons (Nandi and Basumallick, 1991).

4.3 PRINCIPAL COMPONENTS ANALYSIS (PCA)

In the whole molecule approaches considered hitherto, the relevant properties were assumed to have been identified; the burden of the discussion was then how best to calculate these properties and then test the resulting $\log K_{ow}$ predictions. In contrast, PCA does not make this presupposition and attempts to identify the more important variables.

The method known as factor analysis is an example of PCA (Cramer, 1980: see Chapter 2). This approach is rather coarse-grained. A somewhat more focused attempt was presented by Dunn *et al.* (1986b). The first, most important, factor turned out to be a size-related solute parameter (surface area, volume). The second appeared to be of electronic origin (e.g., acidity). However useful PCA may be in other areas of chemistry, for predicting K_{ow} it is hopelessly outclassed at the present time by the fragment methods already considered (Taylor, 1990).

4.4 NEURAL NETWORKS (CONNECTIONISM, PARALLEL DISTRIBUTED PROCESSING

Neural networks is a recently elaborated field of study and the subject of intense current interest. Jubak (1992) provides a non-technical account, while short technical introductions have been given by Hinton (1992) and Lippmann (1987).

In the human brain, neurons collect signals from many others and send out signals to equally numerous receiving neurons. Artificial neural networks attempt to model this brain structure (since our knowledge of how the brain works is still rudimentary, such an attempt is simplistic and elementary). In artificial neural networks, a large number of model neurons ('units' represented by a computer program) are interconnected, and the signals between units are weighted according to chosen initial conditions and subsequent learning procedures (Fig. 5.1).

Neural networks are an outgrowth of artificial intelligence, and have found application in pattern recognition. With respect to K_{ow} prediction, the network is first trained on a large set of compounds of known K_{ow}. For these molecules, a set of descriptors is established, with numerical values. These descriptors are the usual molecular properties calculated by molecular orbital and quantum chemical methods. The system is trained to give the required output (K_{ow}) given the input (descriptor values). The most popular algorithm for training

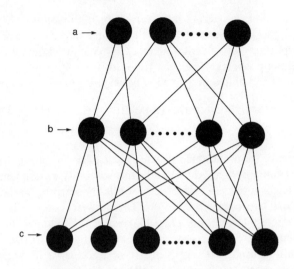

Figure 5.1 Simple neural network (Hinton, 1992)

(i.e., optimizing the inter-neuron weights) is back-propagation. Although neural nets are set up in computer hardware, it should be emphasized that the networks themselves are not von Neumann machines and do not follow a linear sequence program.

Between 8 and 17 parameters have been used in K_{ow} prediction by this method (Bodor *et al.*, 1994; Cense *et al.*, 1994, 1995; Grunenberg and Herges, 1995; Zhang and Liu, 1995). Related quantities, such as water solubility and RP-HPLC retention times, have been similarly predicted (Bodor *et al.*, 1991; Cupid *et al.*, 1993). The predictive accuracy is about the same as that of ordinary regression analysis (Bodor *et al.*, 1994; Grunenberg and Herges, 1995). The method of neural networks is thus only of passing interest at this time.

5 LINEAR SOLVATION ENERGY RELATIONSHIPS (LSER)

This is an example of a whole molecule approach of such moment as to deserve a separate consideration.

The LSER method was summarized by Taft *et al.* (1985). It is a generalized treatment of solvation effects which assumes that solute–solvent interactions are frequently of two kinds: (a) non-specific (dipolarity/polarizability) and (b) hydrogen-bond formation. The effects in (b) are considered to comprise hydrogen bond acceptor and donor kinds, of both solute and solvent. For transfer properties, an additional term for the formation of the solute cavity in the solvent was included; this term depends upon the volume of the solute molecule and the energy of separation of solvent molecules to accommodate the solute.

Thus for a solution property Z, either for a single solute in a series of solvents or for a series of solutes in a fixed solvent, the general relation

$$Z = Z_0 + \text{(cavity formation term)} + \text{(dipolarity/polarizability)}$$
$$+ \text{(hydrogen bond terms)} \quad (5.9)$$

where Z_0 is the property for a reference system. The cavity formation system is scaled by a molar volume and δ_H, the Hildebrand solubility parameter of the solvent. The other terms are scaled by π^* (for solute dipolarity/polarizability) and α and β (for hydrogen bonding). These were originally called *solvatochromic* parameters (Kamlet *et al.*, 1983), since the defining molecular properties were often spectroscopic solvent shifts (among other properties).

The general relation, Eq. (5.9), proved to be successful in correlating quantitatively a large number of solvation effects. Among these are: NMR shifts (^{15}N, ^{13}C, ^{19}F, ^{77}Se, ^{1}H), ESR spectra, thermodynamic transfer quantities, substitution reaction rates, complex formation and association constants, etc. Taft *et al.* (1985) provide a detailed bibliography.

The application of Eq. (5.9) to K_{ow} prediction requires special attention to solute hydrogen-bond acidity and basicity (Abraham, 1993). The Taft–Kamlet–Abraham group has made exemplary contributions to the LSER correlation method (Abraham *et al.*, 1994a,b; Kamlet *et al.*, 1984, 1988a,b). Others have considered K_{ow} and LSER in wider or different contexts (El Tayar *et al.*, 1991; Famini *et al.*, 1992; Herndon, 1993; Marcus, 1991, 1992). Leahy (1986) substituted a van der Waals molecular volume for the original bulk molar volume. Abraham (1993) introduced an 'excess molar refraction' contribution to π^*. Meyer and Maurer (1993, 1995) obtained a general correlation for 826 partition coefficients, representing 101 solutes in 20 organic/solvent systems at 25°C.

REFERENCES

Abraham, M. H. (1993) *Chem. Soc. Rev.* **22**, 73–83.
Abraham, M.H., Chadha, H. S., Dixon, J.P. and Leo, A. J. (1994a) *J. Phys. Org. Chem.* **7**, 712–16.
Abraham, M. H., Chadha, H. S., Whiting, G. S. and Mitchell, R. C. (1994b) *J. Pharm. Sci.* **83**, 1085–1100.
Alkorta, I. and Villar, H. O. (1992) *Int. J. Quantum Chem.* **44**, 203–18.
Arbuckle, W. B. (1983) *Environ. Sci. Technol.* **17**, 537–42.
Banerjee, S. and Howard, P. H. (1988) *Environ. Sci. Technol.* **22**, 839–41.
Barton, A. F. M. (1983) *CRC Handbook of Solubility Parameters and Other Cohesion Parameters*, CRC Press, Boca Raton.

Bezdek, J. C. and Pal, S. K., eds. (1992) *Fuzzy Models for Pattern Recognition*, IEEE Press, New York.
Bodor, N. and Huang, M.-J. (1991) *J. Comput. Chem.* **12**, 1182–6.
Bodor, N. and Huang, M.-J. (1992) *J. Pharm. Sci.* **81**, 272–81.
Bodor N., Gabanyi, Z. and Wong, C.-K. (1989) *J. Am. Chem. Soc.* **111**, 3783–6.
Bodor, N., Harget, A. and Huang, M.-J. (1991) *J. Am. Chem. Soc.* **113**, 9480–3.
Bodor, N., Huang, M.-J. and Harget, A. (1994) *J. Mol. Struct.* **309**, 259–66.
Brinck, T., Murray, J. S. and Politzer, P. (1993) *J. Org. Chem.* **58**, 7070–3.
Broto, P., Moreau, G. and Vandycke, C. (1984) *Eur. J. Med. Chem. -Chim. Ther.* **19**, 71–8.
Calvino, R., Gasco, A. and Leo, A. (1992) *J. Chem. Soc. Perkin Trans.* 2, 1643–6.
Camilleri, P., Watts, S. A. and Boraston, J. A. (1988) *J. Chem. Soc. Perkin Trans.* 2, 1699–1707.
Campbell, J. R. and Luthy, R. G. (1985) *Environ. Sci. Technol.* **19**, 980–5.
Cense, J. M., Diawawa, B., Legendre, J. J. and Roullet, G. (1994) *Chemom. Intell. Lab. Sys.* **23**, 301–8.
Cense, J. M., Diawara, B., Legendre, J. J. and Roullet, G. (1995) *AIP Conf. Proc.* **330**, 556–61.
Chastrette, M., Tiyal, F. and Peyrand, J.-F. (1990) *C. R. Acad. Sci. Paris, Ser. II* **311**, 1057–60.
Cramer, R. D. (1980) *J. Am. Chem. Soc.* **102, 1837–59.**
Cupid, B. C., Nicholson, J. K., Davis, P., *et al.* (1993) *Chromatographia* **37**, 241–9.
Dallos, A., Wienke, G., Ilchmann, A. and Gm.ehling, J. (1993) *Chem.-Ing.-Tech.* **65**, 201–3.
Doucette, W. J. and Andren, A. W. (1987) *Environ. Sci. Technol.* **21**, 821–4.
Doucette, W. J. and Andren, A. W. (1988) *Chemosphere* **17**, 345–59.
Dunn, W. J., Block, J. H. and Pearlman, R. S. (eds) (1986a) *Partition Coefficient Determination and Estimation*, Pergamon, New York.
Dunn, W. J., Grigoras, S. and Johansson, E. (1986b) In Dunn *et al.* (1986a), pp. 21–35.
Dunn, W. J., Koehler, M. G. and Grigoras, S. (1987) *J. Med. Chem.* **30**, 1121–6.
El Tayar, N., Tsai, R.-S., Testa, B., Carrupt, P.-A. and Leo, A. (1991) *J. Pharm. Sci.* **80**, 590–8.
Famini, G. R., Penski, C. A. and Wilson, L. Y. (1992) *J. Phys. Org. Chem.* **5**, 395–408.
Finizio, A., DiGuardo, A. and Vighi, M. (1994) *SAR QSAR Environ. Res.* **2**, 249–60.
Finizio, A., Sicbaldi, F. and Vighi, M. (1995) *SAR QSAR Environ. Res.* **3**, 71–80.
Fredenslund, A., Jones, R. L. and Prausnitz, J. M. (1975) *A. I. Ch. E. J.* **21**, 1086–99.
Fujita, T. (1983) *Prog. Phys. Org. Chem.* **14**, 75–113.
Fujita, T., Iwasa, J. and Hansch, C. (1964) *J. Am. Chem. Soc.* **86**, 5175–80.
Ghose, A. K. and Crippen, G. M. (1986) *J. Comput. Chem.* **7**, 565–77.
Ghose, A. K., Pritchett, A. and Crippen, G. M. (1988) *J. Comput. Chem.* **9**, 80–90.
Grunenberg, J. and Herges, R. (1995) *J. Chem. Inf. Comput. Sci.* **35**, 905–11.
Güsten, H. and Sabljić, A. (1993) *Polycyclic Aromat. Compd.* **3**(Suppl.), 267–76.
Güsten, H., Horvatić, D. and Sabljić, A. (1991) *Chemosphere* **23**, 199–213.
Hansch, C. (gen. ed.) (1990) *Comprehensive Medicinal Chemistry*, 6 vols., Pergamon, Oxford.
Hansch, C. and Leo, A. (1979) *Substituent Constants for Correlation Analysis in Chemistry and Biology*, Wiley-Interscience, New York.
Hansch, C. and Leo, A. J. (1995) *Exploring QSAR: fundamentals and applications in chemistry and biology*, American Chemical Society, Washington.
Herndon, W. C. (1993) *J. Phys. Org. Chem.* **6**, 634–6.
Hinton, G. E. (1992) *Scientific American* **267**(3), 144–51.

Hopfinger, A. J. and Battershell, R. D. (1976) *J. Med. Chem.* **19**, 569–73.

Iwase, K., Komatsu, K., Hirono, S., Nakagawa, S. and Moriguchi, I. (1985) *Chem. Pharm. Bull.* **33**, 2114–21.

Jubak, J. (1992) *In the Image of the Brain*, Little, Brown and Company, Boston.

Kamlet, M. J., Abboud, J.-L. M., Abraham, M. H. and Taft, R. W. (1983) *J. Org. Chem.* **48**, 2877–87.

Kamlet, M. J., Abraham, M. H., Doherty, R. M. and Taft, R. W. (1984) *J. Am. Chem. Soc.* **106**, 464–6.

Kamlet, M. J., Doherty, R. M., Abraham, M. H., Marcus, Y. and Taft, R. W. (1988a) *J. Phys. Chem.* **92**, 5244–55.

Kamlet, M. J., Doherty, R. M., Carr, P. W., Mackay, D., Abraham, M. H. and Taft, R. W. (1988b) *Environ. Sci. Technol.* **22**, 503–9.

Kantola, A., Villar, H. O. and Loew, G. H. (1991) *J. Comput. Chem.* **12**, 681–9.

Kasai, K., Umeyama, H. and Tomonaga, A. (1988) *Bull. Chem. Soc. Jpn.* **61**, 2701–6.

Katagi, T., Miyakado, M., Takayama, C. and Tanaka, S. (1995) *ACS Symp. Ser.* **606**, 48–61.

Kier, L. B. (1980) In *Physical Chemical Properties of Drugs*, (ed.) S. H. Yalkowsky, A. A. Sinkula and S. C. Valvani, Marcel Dekker, New York, pp. 277–319.

Kier, L. B. and Hall, L. H. *Molecular Connectivity in Chemistry and Drug Research*, Academic Press, New York.

Kim, K. H. (1995) *Quant. Struct.-Act. Relat.* **14**, 8–18.

Kim, K. H. and Kim, D. H. (1995) *Bioorg. Med. Chem.* **3**, 1389–96.

Klopman, G. and Iroff, L. D. (1981) *J. Comput. Chem.* **2**, 157–60.

Klopman, G. and Wang, S. (1991) *J. Comput. Chem.* **8**, 1025–32.

Klopman, G., Li, J.-Y., Wang, S. and Dimayuga, M. (1994) *J. Chem. Inf. Comput. Sci.* **34**, 752–81.

Klopman G., Namboodiri, K. and Schochet, M. (1985) *J. Comput. Chem.* **6**, 28–38.

Koehler, M. G., Grigoras, S. and Dunn, W. J. (1988) *Quant. Struct.-Act. Relat.* **7**, 150–9.

Leahy, D. E. (1986) *J. Pharm. Sci.* **75**, 629–36.

Leo, A. (1983) *J. Chem. Soc. Perkin Trans.* 2, 825–38.

Leo, A. J. (1990) in Hansch (1990), vol. 4, Chapter 18.7.

Leo, A. J. (1991) *Methods in Enzymology* **202**, 544–91.

Leo, A. J. (1993) *Chem. Rev.* **93**, 1281–1306.

Leo, A., Hansch, C. and Elkins, D. (1971) *Chem. Rev.* **71**, 525–616.

Lippmann, R. P. (1987) *IEEE ASSP MAG.* (April), 4–22. Also reprinted in Bezdek and Pal (1992), pp. 417–35.

Lukovits, I. (1990) *Quant. Struct.-Act. Relat.* **9**, 277–31.

Lyman, W. J., Reehl, W. F. and Rosenblatt, D. H. (1982) *Handbook of Chemical Property Estimation Methods*, McGraw-Hill, New York.

Makovskaya, V., Dean, J. R., Tomlinson, W. R. and Comber, M. (1995) *Anal. Chim. Acta* **315**, 193–200.

Marcus, Y. (1991) *J. Phys. Chem.* **95**, 8886–91.

Marcus, Y. (1992) *Solvent Extract. Ion Exchange* **10**, 527–38.

Meyer, P. and Maurer, G. (1993) *Ind. Eng. Chem. Res.* **32**, 2105–10.

Meyer, P. and Maurer, G. (1995) *Ind. Eng. Chem. Res.* **34**, 373–81.

Meylan, W. M. and Howard, P. H. (1995) *J. Pharm. Sci.* **84**, 83–92.

Moriguchi, I., Hirono, S., Liu, Q., Nakagome, I. and Matsushita, Y. (1992) *Chem. Pharm. Bull.* **40**, 127–30.

Nandi, N. and Basumallick, I. N. (1991) *Z. Phys. Chem. (München)* **173**, 179–89.

Niemi, G. J., Basak, S. C., Veith, G. D. and Grunwald, G. (1992) *Environ. Toxicol. Chem.* **11**, 893–900.

Nys, G. G. and Rekker, R. F. (1973) *Chim. Ther.* **8**, 521–35.
Nys, G. G. and Rekker, R. F. (1974) *Eur. J. Med. Chem.-Chim. Ther.* **9**, 361–75.
Parker, G. R. (1978) *J. Pharm. Sci.* **67**, 513–16.
Patil, G. S. (1991) *Chemosphere* **22**, 723–38.
Pearlman, R. S. (1986) In Dunn *et al.* (1986a), pp. 3–20.
Pierotti, R. (1976) *Chem. Rev.* **76**, 717–26.
Pyka, A. (1995) *J. Planar Chromatog.* **8**, 52–62.
Reid, R. C., Prausnitz, J. M. and Poling, B. E. (1987) *The Properties of Gases and Liquids*, 4th edition, McGraw-Hill, New York.
Rekker, R. F. (1977) *The Hydrophobic Fragmental Constant*, Elsevier, Amsterdam.
Rekker, R. F. and de Kort, H. M. (1979) *Eur. J. Med. Chem.* **14**, 479–88.
Rekker, R. F. and Mannhold, R. (1992) *Calculation of Drug Lipophilicity*, VCH Verlagsgesellschaft, Weinheim.
Sabljić, A., Güsten, H., Hermens, J. and Opperhuizen, A. (1993) *Environ. Sci. Technol.* **27**, 1394–1402.
Saçan, M. T. and İnel, Y. (1995) *Chemosphere* **30**, 39–51.
Sangster, J. (1993) *LOGKOW — a databank of evaluated octanol–water partition coefficients*, Sangster Research Laboratories, Montreal.
Sasaki, Y., Kubodera, H., Matuszaki, T. and Umeyama, H. (1991) *J. Pharmacobio-Dyn.* **14**, 207–14.
Schüürmann, G. (1995a) *Fresenius Environ. Bull.* **4**, 238–43.
Schüürmann, G. (1995b) *Environ. Toxicol. Chem.* **19**, 2067–76.
Suzuki, T. and Kudo, Y. (1990) *J. Computer-Aided Mol. Des.* **4**, 155–98.
Szász, G., Novák-Hankó, K., Kier, L. B., Hermecz, I. and Kökösi, J. (1983) *Acta Pharm. Hung.* **53**, 195–202.
Taft, R. W., Abboud, J.-L. M., Kamlet, M. J. and Abraham, M. H. (1985) *J. Solution Chem.* **14**, 153–86.
Takács-Novák, K., Nagy, P., Józani, M., Órfi, L., Dunn, W. J. and Szász, G. (1992) *Acta Pharm. Hung.* **61**, 55–64.
Takeuchi, K., Kuroda, C. and Ishida, M. (1990) *J. Chem. Inf. Comput. Sci.* **30**, 22–6.
Taylor, P. J. (1990) In Hansch (1990), Vol. 4, Chapter 18.6.
Viswanadhan, V. N., Ghose, A. K., Revankar, G. R. and Robins, R. K. (1989) *J. Chem. Inf. Comput. Sci.* **29**, 163–72.
Waller, C. L. (1994) *Quant. Struct.-Act. Relat.* **13**, 172–6.
Watarai, H., Tanaka, M. and Suzuki, N. (1982) *Anal. Chem.* **54**, 702–5.
Yalkowsky, S. H. and Valvani, S. C. (1976) *J. Med. Chem.* **19**, 727–8.
Zhang, X.-D. and Liu, Q.-T. (1995) *Gaodeng Xuexiao Huaxue Xuebao* **16**, 1360–3.

CHAPTER 6

Discussion on Log K_{OW} Predictive Methods

In Chapter 5 were described a number of predictive methods for calculating log K_{ow}. Taken as a whole, these methods vary significantly in sophistication, accuracy and applicability. There has, as yet, not appeared any systematic evaluation of all these methods. Discussions by Leo (1990, 1993) and Taylor 1990 cover some of this variety. Comparison of fragment methods will be presented in this chapter in some detail.

The various fragmental methods have in general been perfected to a higher degree than the whole molecule approaches. A number of fragment methods are available as programs for use on a PC; whole molecule methods have not yet been similarly developed. In what follows, the predictive accuracy of some fragmental methods will be compared quantitatively for a series of classes of organic compounds. Identification and classification of these compounds were made with reference to several handbooks: Budavari (1996), Elks and Ganellin (1990) and Negwer (1987) for drugs and medicinal chemicals; Worthing (1987) and Hartley and Kidd (1987) for agrochemicals; Society of Dyes and Colourists (1971) for dyes. Howard and Neal (1992) was helpful for general identification.

1 FRAGMENT METHODS

An early comparison of the methods of Hansch and Leo (1979) and Rekker and De Kort (1979) was presented by Mayer *et al.* (1982). These authors discussed the basic premises and correction factors of both methods; quantitative comparisons were made on 49 simple molecules. They pointed out various inconsistencies in the application of correction factors. At this date, the two methods appeared to be about equally accurate in predicting log K_{ow}.

1.1 COMPUTERIZED VERSIONS

The advantages of using computerized versions of fragmental methods were pointed out by Hansch and Leo (1995) and Leo (1993). This is especially true if the method is at all complicated or is updated at intervals. The calculation of $\log K_{ow}$ of lidocaine by the Rekker method may be an object lesson in this regard. (The Rekker method was chosen as example because it is plainly documented; the principle, however, applies to any such fragment method.)

Table 6.2 shows $\log K_{ow}$ calculated values for lidocaine from several sources. The Rekker method was updated in 1992. The experimental value is 2.26 (Sangster, 1993). The early discrepancy between calculated and experimental values was noted (Rekker and Mannhold, 1992; Rekker et al., 1993) and the rather large difference was attributed to decoupling of resonance between phenyl and NHCO groups due to the ortho substitution steric effect. The first

Table 6.1 Characteristics of computerized fragmental methods (for explanation of entries, see text)

Program	Input	Storage	Formula	Explanation	Exptl. Data	Warnings	Charged compounds	Tautomers
ClogP	SMILES	Y	N(a)	Y	Y	Y	(b)	N
KOWWIN	SMILES	Y	Y(a)	Y	Y	Y	(b)	N
ACD/LogP	draw	Y	Y	Y	Y	Y	(c)	Y
PrologP	draw	Y	Y	Y	N	Y	(d)	N
KlogP	SMILES	Y	N(a)	Y	N	Y	N	N

(a) Program presents two-dimensional structure from SMILES.
(b) Zwitterionic, quaternary compounds and ionized forms of acids and bases.
(c) Zwitterionic compounds.
(d) Zwitterionic compounds and ionized forms of acids and bases (PrologD module).

Table 6.2 Log K_{ow} of lidocaine as calculated by method of Rekker et al.

Entry	log K_{ow} (calcd)	Source	Method Reference
1	3.04	Dross et al. (1992)	Nys and Rekker (1974)
	3.04	Mannhold et al. (1990)	Nys and Rekker (1974)
2	3.33	Rekker and Mannhold (1992)	Rekker and De Kort (1979)
3	3.40	Rekker and Mannhold (1992)	Rekker and Mannhold (1992)
4	2.30	Rekker et al. (1993)	Rekker and De Kort (1979) with steric correction
	2.31	Mannhold et al. (1995)	Rekker and Mannhold (1992)
5	3.13	PrologP 5.1	—

four entries in Table 6.2 were presumably the result of manual calculation; a generally available computerized version (PrologP 5.1) gives a result different from all the rest.

There are a number of fragmental methods which are available in computerized form; features of these PC programs are given in Table 6.1. The various entries in this table are described here.

Input indicates how the user tells the program which molecule(s) is to be calculated. SMILES is the acronym for Simplified Molecular Input Line Entry System. It is a method (Weininger, 1988) for representing the structure of the molecule as a string of ASCII symbols. 'Draw' means that the user draws the two-dimensional structure by mouse manipulation.

Storage indicates whether or not the molecular structures can be stored in a file for later retrieval and calculation by the user.

Formula indicates whether or not the molecular formula corresponding to the user's input is shown. The formula is a convenient check for ascertaining that the input corresponds to the molecule the user wishes to calculate.

Explanation indicates whether the program provides a summary of the various fragment contributions, correction factors, etc. used in the calculation.

Experimental data indicates whether or not the program provides measured $\log K_{ow}$ data from the literature for the molecule under consideration. (The choice and number of data presented, in those cases where this feature is present, varies from program to program).

Warnings refers to messages given to the user concerning the input and the molecule corresponding to it, such as: correctness of SMILES entry, observance of usual atomic valences, etc. ClogP and ACD/LogP alert the user to calculated data which are unrealistically high; KlogP includes a user-specified cut-off limit for this purpose.

Charged compounds indicate what ionic species, if any, are calculated by the program.

Tautomers indicates whether or not the program alerts the user to the possible existence of tautomeric forms of the molecule to be calculated. ACD/LogP is the only product which has this feature, and it calculates $\log K_{ow}$ for each tautomer, as desired.

1.2 THE PREDICTIVE PERFORMANCE OF FRAGMENT METHODS

An extensive evaluation of several estimation methods was reported by Kühne *et al* (1994) and Schüürmann *et al* (1995a). They used 650 compounds; these contained elements C, H, halogen, N, O, P and S. Both simple and complex molecules (drugs, agrochemicals, PCBs) were represented. The results were presented only in statistical form. Their data have been re-worked here and appear in Table 6.3.

Table 6.3. RMS deviations for calculated log of simple compounds (number of compounds in parentheses)

Elements in compounds (log K_{ow} range)	Hansch and Leo (a,b)	Meylan and Howard (c)	ACD (d)	Rekker (e)	Suzuki and Kudo (a)	Ghose and Crippen (a)	Broto (a)	Klopman (f)
C, H (0.37, 6.50)	0.25 (73)	0.40 (73)	0.35 (73)	0.40 (73)	0.28 (73)	0.50 (73)	0.59 (73)	0.28 (73)
C, H, halogen (0.51, 5.04)	0.19 (76)	0.27 (76)	0.29 (76)	0.29 76)	0.41 (76)	0.39 (76)	0.55 (71)	0.43 (76)
C, H, O (−1.76, 7.45)	0.22 (167)	0.25 (167	0.22 (167)	0.36 (167)	0.41 (167)	0.43 (165)	0.36 (158)	0.34 (167)
C, H, O, halogen (0.22, 5.12)	0.23 (27)	0.24 (27)	0.20 (27)	0.28 (27)	0.61 (27)	0.28 (27)	0.36 (27)	0.33 (27)
C, H, N (-2.04, 3.82)	0.17 (68)	0.25 (68)	0.17 (68)	0.27 (68)	0.29 (68)	0.57 (65)	0.40 (57)	0.44 (68)
C, H, N, O (−2.11, 3.19)	0.10 (72)	0.20 (72)	0.20 (72)	0.43 (72)	0.51 (72)	0.45 (69)	0.59 (70)	0.40 (72)
C, H, N, halogen (1.88, 3.69)	0.13 (11)	0.36 (11)	0.53 (11)	0.30 (11)	0.34 (11)	0.34 (11)	0.44 (10)	0.34 (11)
C, H, N, O, halogen (0.77, 4.22)	0.26 (13)	0.33 (13)	0.52 (13)	0.61 (13)	1.01 (13)	0.49 (13)	0.70 (12)	0.76 (13)
C, H, S, other (−1.35, 4.15)	0.42 (13)	0.18 (13)	0.17 (12)	0.45 (13)	1.66 (13)	0.96 (12)	0.47 (11)	0.83 (13)

(a) Schüürmann *et al.* (1995a) and private communication.
(b) ClogP 4.34.
(c) KOWWIN 1.53.
(d) ACD/LogP 1.0
(e) PrologP 5.1.
(f) KlogP contained in ToxAlert 1.2.

In this table, results of calculations on simple molecules only are shown (G. Schüümann, private communication). The errors are reported as simple root-mean-square deviations for the methods and classes of compounds shown. For the methods of Hansch and Leo, Suzuki and Kudo, Ghose and Crippen and that of Broto, Schüümann's detailed results were used; for the remaining methods, recent computer versions supplied the data indicated.

Readers are cautioned not to draw excessively far-reaching conclusions from the information in Table 6.3, particularly for groups of small sample size. When the sample is small, 'embarrassing' cases may be over- or under-represented, with consequent skewing of the overall statistics. In general, however, the methods of Hansch and Leo, Meylan and Howard and ACD performed better than the others.

1.3 TESTS ON DRUG MOLECULES

In a series of investigations, the Rekker group has examined the predictive accuracy of several methods for representative drugs (Dross *et al.*, 1992; Mannhold *et al.*, 1990, 1995; Rekker and Mannhold, 1992; Rekker *et al.*, 1993). In 1995 it was concluded (Mannhold *et al.*, 1995) that the Hansch and Leo and the Rekker methods were generally more accurate than the others (these studies did not include the Meyland and Howard and the ACD methods). A similar comparison was made by Moriguchi *et al.* (1992) on a smaller set of drug molecules.

A quantitative test of several computerized fragment methods is presented in Table 6.4, comprising 88 drugs of various kinds.

1.4 TESTS ON OTHER COMPLICATED MOLECULES

In general, all methods performed better on simple, rather than complicated, molecules. This is not surprising. The results for other complicated molecules may be ascertained by inspection of Tables 6.5–6.12. In many of these tables, it can be seen readily which methods take into account unusual aspects of the molecules concerned.

Although some methods are better than others, all fail in the cases of organic dyes and radiopaque compounds (Tables 6.11 and 6.12). These compounds are a calculator's nightmare. Partition coefficients of dye molecules have not even been measured accurately.

If the molecule can exist in two or more tautomeric forms, the partition coefficient, as measured, will be the result of the predominant forms in octanol and water (these may not be the same). A well-designed calculational method can distinguish the partition coefficients of different tautomers, which can be very different from each other (Hansch and Leo, 1995; Leo, 1990; Taylor,

Table 6.4. Calculated log K_{ow} of drugs and medicinal chemicals (neutral molecules unless otherwise indicated)

Formula	Name	CAS registry number	K_{ow} expt. (a)	Hansch and Leo (b)	Meyland and Howard (c)	ACD (d)	Rekker (e)/(f)	Klopman (g)
Analgesic								
$C_{18}H_{21}NO_3$	codeine	76-57-3	1.14R	0.82	1.28	1.83	2.46/-	3.02
$C_{14}H_{10}F_3NO_2$	flufenamic acid	530–78–9	5.25R	5.51	5.15	5.62	4.52/5.81	4.62
$C_{19}H_{16}ClNO_4$	indomethacin	53-86-1	4.27R	4.18	4.23	3.10	2.92/-	4.75
$C_{21}H_{27}NO$	methadone	76-99-3	3.93R	3.21	4.17	4.20	3.70/-	5.18
$C_{17}H_{19}NO_3$	morphine	57-27-2	0.76R	0.24	0.72	1.06	1.86/-	2.85
$C_{19}H_{20}N_2O_2$	phenylbutazone	50-33-9	3.16R	3.16	3.52	3.47	2.66/-	2.91
Antianginal								
$C_{17}H_{18}N_2O_6$	nifedipine	21829-25-4	–	2.35	2.50	3.05,1.93	2.67/1.96	3.35
Antiarrhythmic								
$C_{25}H_{29}I_2NO_3$	amiodarone	1951-25-3	–	8.75	8.81	8.58	9.84,9.40	8.55
$C_{22}H_{30}N_2$	aprinidine	37640-71-4	4.86	4.79	5.90	5.77	5.65/—	5.42
$C_{27}H_{31}NO_3$	asocainol	77400-65-8	4.85	5.80	6.03	6.20	6.53/—	6.22
$C_{18}H_{25}N_3O_5$	carocainide	66203-00-7	1.38R	1.98	2.01	1.60	2.84/2.65	2.08
$C_{22}H_{26}N_2O_4S$	diltiazem	42399-41-7	2.70R	3.57	2.79	4.53	3.79/4.53	4.14
$C_{21}H_{29}N_3O$	disopyramide	3737-09-5	2.58	1.57	2.96	2.86	2.39/2.59	4.23
$C_{11}H_{17}NO$	mexiletine	31828-71-4	2.15R	2.57	2.61	2.16	3.13/2.93	2.13
$C_{22}H_{25}N_3O_4S$	moricizine	31883-05-3	2.98R	3.25	1.97	2.67	2.48/2.82	3.59
$C_{21}H_{27}N_3O_3$	nicainoprol	76252-06-7	1.63R	1.38	1.69	1.75	1.90/1.44	2.23
$C_{13}H_{21}N_3O$	procainamide	51-06-9	0.88R	1.24	0.97	1.23	1.13/1.13	1.70
$C_{21}H_{27}NO_3$	propasenone	54063-53-5	4.63	3.42	3.37	4.63	4.11/3.72	3.72
$C_{20}H_{24}N_2O_2$	quinidine	56-54-2	2.64R	2.93	3.29	3.36	3.15/2.76	3.23
$C_{12}H_{20}N_2O_3S$	sotalol	3930-20-9	0.24R	0.23	0.37	0.32	0.62/0.33	1.11
$C_{27}H_{38}N_2O_4$	verapamil	52-53-9	3.83R	3.79	4.80	5.03	5.92/6.16	6.19

(Continued)

	CAS						
Antibacterial							
$C_{16}H_{19}N_3O_4S$ ampicillin (h)	69-53-4	-1.13R	-1.27	-0.88	n.c.	1.07(i)/0.93	n.c.
$C_{11}H_{12}Cl_2N_2O_5$ chloroamphenicol	56-75-7	1.14R	0.69	0.92	1.02	0.69/0.32	1.06
$C_{14}H_{18}N_4O_3$ trimethoprim	738-70-5	0.91R	0.48	0.73	0.79	2.01/-0.07	1.38
Anticholinergic							
$C_{17}H_{23}NO_3$ atropine	51-55-8	1.83R	1.32	1.91	1.53	2.12/1.88	2.24
$C_{20}H_{23}NS$ methixene	4969-02-2	–	5.86	5.67	5.46	5.61/–	5.51
Anticonvulsant							
$C_{15}H_{12}N_2O_2$ phenytoin	57-41-0	2.47R	2.09	2.16	2.52	1.30/2.76	1.80
Antidepressant							
$C_{17}H_{18}F_3NO$ fluoxetine	54910-89-3	–	4.05	4.65	4.35	4.63/–	4.78
$C_{19}H_{24}N_2$ imipramine	50-49-7	4.80R	4.49	5.01	4.47	4.53/4.43	4.29
Antidiarrhoeal							
$C_{29}H_{33}ClN_2O-2$ loperamide	53179-11-6	–	3.90	5.15	5.08	3.63/4.43	4.29
Antiemtic							
$C_{16}H_{21}N_5O_2$ alizapride	59338-93-1	1.79	2.74	1.80	0.77,178	1.84/2.28	1.83
$C_{17}H_{27}N_3O_4S$ amisulpride	71675-85-9	1.10	1.73	1.11	1.29	0.52/0.71	0.57
$C_{17}H_{19}ClN_2S$ sulpiride	15676-16-1	0.62R	1.11	0.65	0.45	0.68/1.21	0.93
$C_{22}H_{29}N_3S_2$ thiethylperazine	58-73-1	3.40R	6.31	5.24	5.05	4.88/5.67	5.46
Antihistaminic							
$C_{10}H_{16}N_6S$ cimetidine	51481-61-9	-0.40R	0.35	0.57	0.36	-0.30/0.63	0.36
$C_{17}H_{21}NO$ diphenhydramine	58-73-1	3.40R	3.36	3.11	3.66	3.29/3.41	3.92

(Continued)

Table 6.4. (Continued)

Formula	Name	CAS registry number	K_{ow} expt. (a)	Hansch and Leo (b)	Meyland and Howard (c)	ACD (d)	Rekker (e)/(f)	Klopman (g)
Antihyperlipoproteinaemic								
$C_{12}H_{15}ClO_3$	clofibrate	637-07-0	—	3.68	3.62	3.32	4.02/4.21	3.40
Antihypertensive								
$C_9H_{15}NO_3S$	captopril	62571-86-2	—	1.02	0.84	1.51	1.02/0.89	0.87
$C_7H_6ClN_3O_4S_2$	chlorothiazide	58-94-6	−0.24R	−0.41	−0.23	−0.18 − 0.63	−0.10/—	−0.16
$C_{19}H_{25}N_5O_4$	terazosin	63590-64-7	−0.38R	2.47	1.47	−0.64	1.91/2.45	0.59
Antipsychotic								
$C_{21}H_{23}ClFNO_2$	haloperidol	52-86-8	3.36R	3.52	4.20	4.06	3.38/3.57	4.72
β-blockers								
$C_{18}H_{28}N_2O_4$	acebutolol	37517-30-9	1.77R	1.63	1.19	2.59	2.08/1.51	1.67
$C_{15}H_{23}NO_2$	alprenolol	13655-52-2	3.10R	2.65	2.81	2.88	3.54/2.89	2.67
$C_{14}H_{22}N_2O_3$	atenolol	29122-68-7	0.16R	−0.11	−0.03	0.10	0.71/0.05	0.90
$C_{14}H_{20}N_2O_2$	bunitrolol	34915-68-9	2.00R	1.74	1.42	1.60	2.01/1.69	1.77
$C_{17}H_{27}NO_4$	metipranolol	22664-55-7	2.66	2.55	2.66	2.67	3.32/2.84	2.71
$C_{15}H_{25}NO_3$	metoprolol	37350-58-6	1.88R	1.20	1.69	1.76	2.32/1.76	2.22
$C_{15}H_{23}NO_3$	oxprenolol	6452-71-7	2.18R	1.69	1.83	2.29	3.09/2.44	2.26
$C_{18}H_{29}NO_2$	penbutolol	38363-40-5	4.15R	4.04	4.20	4.17	4.63/4.23	3.31
$C_{16}H_{21}NO_2I$	propanolol	525-66-6	3.09R	2.75	2.60	3.10	3.62/3.03	2.75
Cardiotonic								
$C_{12}H_{12}N_2O_2S$	enoximone	77671-31-9	—	2.26	1.86	2.43	1.57/1.73	1.89
$C_7H_8N_4O_2$	theophylline	58-55-9	−0.02R	−0.06	−0.39	0.06	−2.03/—	−0.63
Diuretic								
$C_{13}H_{12}Cl_2O_4$	ethacrynic acid	58-54-8	—	3.36	3.69	3.38	3.81/3.88	3.88
$C_{12}H_{11}ClN_2O_5S$	furosemide	54-31-9	2.03R	1.77	2.32	2.92	1.31/1.38	2.17

(Continued)

Dopamine antagonist

$C_{17}H_{25}N_3O_5S$	veralipride	66644-81-3	1.47	0.57	0.57	1.30	0.90/1.38	1.14

Local anaesthetic

$C_{14}H_{22}N_2O$	lidocaine	137-58-6	2.26R	2.06	1.66	2.36	3.13/2.31	2.67
$C_{15}H_{24}N_2O_2$	tetracaine	94-24-6	3.73R	3.65	3.02	3.75	3.41/3.55	3.67

Narcotic

$C_{21}H_{23}NO_5$	heroin	561-27-3	–	1.15	1.80	2.14	2.71/–	3.68

Nootropic

$C_6H_{10}N_2O_2$	piracetam	7491-74-9	–1.54R	–1.49	–1.40	–0.47	–1.91/–1.96	–1.20

Psychotropic/hallucinogenic

$C_{21}H_{26}O_2$	cannabinol	521-35-7	–	7.39	7.23	7.35	6.84/–	6.68
$C_{20}H_{25}N_3O$	LSD-25	50-37-3	2.95R	2.49	2.26	2.67	2.99/–	3.52
$C_{11}H_{17}NO_3$	mescaline	54-04-6	–	0.18	0.85	0.90	1.66/–	1.27
$C_{15}H_{24}N_2O_4S$	tiapride	51012-32-9	0.90R	1.10	0.53	0.99	0.88/–	0.93

Sedative

$C_{14}H_{22}BrN_3O_2$	bromopride	4093-35-0	2.83R	2.23	1.94	2.77	2.15/2.58	2.52
$C_{12}H_{12}N_2O_3$	phenobarbitol	50-06-6	1.47R	1.37	1.33	1.71	–0.60/1.23	1.44

Stimulant

$C_8H_{10}N_4O_2$	caffeine	58-08-2	–0.07R	–0.06	0.16	–0.07	–1.80/–	–0.76
$C_{17}H_{21}NO_4$	cocaine	50-36-2	2.30R	2.72	2.17	3.01	3.42/–	2.45
$C_{10}H_{14}N_2$	nicotine	54–11-5	1.17R	1.32	1.00	0.72	0.91/–	1.60

(Continued)

Table 6.4. (*Continued*)

Formula	Name		CAS registry number	K_{ow} expt. (a)	Hansch and Leo (b)	Meyland and Howard (c)	ACD (d)	Rekker (e)/(f)	Klopman (g)
Tranquillizer									
$C_{16}H_{13}ClN_2O$	diazepam		439-14-5	2.82R	3.08	2.70	2.15,2.96	2.92/3.18	3.39
$C_{22}H_{26}F_3N_3OS$	fluphenazine		69-23-8	4.36R	5.91	4.13	4.84	4.34/4.86	4.74
$C_{18}H_{19}F_3N_2S$	fluopromazine		146-54-3	5.19R	5.63	5.52	5.70	5.12/5.40	5.55
$C_{19}H_{24}N_2OS$	levomepramazine		60-99-1	4.68R	4.80	5.06	5.05	4.56/–	4.76
$C_9H_{18}N_2O_4$	meprobamate		57-53-4	0.70R	0.28	0.98	0.70	0.85/–	0.96
$C_{20}H_{25}N_3S$	perazine		84-97-9	–	5.22	4.15	4.04	3.77/4.53	4.19
$C_{21}H_{26}ClN_3OS$	perphenazine		58-39-9	4.20R	5.58	3.82	4.49	3.93/4.57	4.32
$C_{17}H_{20}N_2S$	promethazine		60-87-7	4.75	4.73	4.49	4.69	4.55/4.60	4.37
$C_{17}H_{20}N_2S$	promazine		58-40-2	4.55R	4.36	4.56	4.63	3.97/4.38	4.47
$C_{17}H_{26}N_2O_4S$	sultopride		53583-79-2	1.06	1.93	1.33	1.16	1.56/1.82	1.87
$C_{21}H_{26}N_2S_2$	thioridazine		50-52-2	5.90R	6.42	6.45	6.13	5.76/5.92	5.70
$C_{21}H_{24}F_3N_3S$	trifluoperazine		117-89-5	5.03R	6.99	5.11	5.11	4.91/5.56	5.27
$C_{18}H_{22}N_2S$	trimeprazine		84-96-8	4.81	4.75	4.98	4.98	4.49/4.90	4.78
Vaso- and bronchodilator									
$C_{23}H_{25}N$	fendiline		13042-18-7	–	5.73	5.76	6.00	6.11/6.05	6.13
$C_{17}H_{21}NO_4$	fenoterol		13392-18-2	–	0.83	1.22	0.89	1.78/1.69	2.79
$C_{20}H_{21}NO_4$	papaverine		58-74-2	–	3.02	3.69	3.42	4.38/4.51	4.17
$C_{14}H_{20}N_2O_2$	pindolol		13523-86-9	1.75R	1.67	1.48	1.97	2.51/1.74	1.91

n.c.=not calculable.
(a) R designates Recommended Value from Sangster (1993). Others are ion-corrected data from Mannhold *et al.* (1990).
(b) ClogP for Windows 1.0.0.
(c) KOWWIN 1.53
(d) ACD/LogP 1.0. Where two values are given, these refer to separate tautomers.
(e) PrologP 5.1.
(f) Data in italics from Mannhold *et al.* (1995); others from Rekker and Mannhold (1992).
(g) KlogP contained in ToxAlert 1.2.
(h) Zwitterionic compound.
(i) PrologD 3.0

Table 6.5. Calculated log K_{ow} of crown ethers

Formula	Name	CAS registry number	K_{ow} expt. (a)	Hansch and Leo (b)	Meyland and Howard (c)	ACD (d)	Rekker (e)	Klopman (f)
Benzo crown ethers								
$C_{14}H_{20}O_5$	15-crown-5	14098-44-3	0.91	1.11	1.22	1.12	2.48	2.89
$C_{16}H_{24}O_6$	18-crown-6	14098-24-9	0.58	0.87	0.94	0.99	2.51	3.15
$C_{18}H_{28}O_7$	21-crown-7	67950-78-1	0.57	0.62	0.67	0.85	2.55	3.41
$C_{20}H_{32}O_8$	24-crown-8	72216-45-6	0.45	0.37	0.40	0.72	2.58	3.67
$C_{22}H_{36}O_9$	27-crown-9	63144-76-3	0.23	0.19	0.12	0.58	2.62	3.94
$C_{24}H_{40}O_{10}$	30-crown-10	77963-62-5	0.03	-0.12	-0.15	0.45	2.65	4.20
$C_{26}H_{44}O_{11}$	33-crown-11	104946-62-5	-0.09	(g)	-0.43	0.31	2.69	4.46
Dibenzo crown ethers								
$C_{20}H_{24}O_6$	[3,3]18-crown-6	14187-32-7	2.20	3.21	3.31	2.79	4.82	4.83
$C_{24}H_{32}O_8$	[4,4]24-crown-8	14174-09-5	2.11	2.71	2.76	2.52	4.89	5.30
$C_{26}H_{36}O_9$	[2,7]27-crown-9	104946-50-1	1.63	2.53	2.48	2.74	4.92	5.57
$C_{28}H_{40}O_{10}$	[5,5]30-crown-10	17455-25-3	1.80	2.22	2.21	2.25	4.96	5.83
$C_{28}H_{40}O_{10}$	[2,8]30-crown-10	104946-54-5	1.82	2.22	2.21	2.60	4.96	5.83
$C_{30}H_{44}O_{11}$	[4,7]33-crown-11	87586-46-7	1.45	(g)	1.94	2.11	4.99	6.09
$C_{40}H_{64}O_{16}$	[8,8]48-crown-16	121284-20-6	0.52	(g)	0.05	1.43	5.17	7.40
4'-(tert-butyl) benzo crown ethers								
$C_{26}H_{44}O_9$	27-crown-9	121445-65-6	1.89	1.95	2.03	2.27	4.70	5.48
$C_{28}H_{48}O_{10}$	30-crown-10	121445-66-7	1.69	1.71	1.76	2.13	4.73	5.75

(a) Recommended Values (Sangster, 1993).
(b) ClogP for Windows 1.0.0.
(c) KOWWIN 1.53.
(d) ACD/LogP 1.10.
(e) PrologP 5.1.
(f) KlogP contained in ToxAlert 1.2
(g) SMILES input not accepted by program.

Table 6.6. Calculated log K_{OW} of amino acids and unprotected peptides

Formula	Name	CAS registry number	log K_{ow} expt. (a)	Hansch and Leo (b)	Meyland and Howard (c)	ACD (d)	Rekker (e)
$C_3H_7NO_2$	alanine	56-41-7	-2.85	-3.12	-2.99	-2.77	-2.74
$C_2H_5NO_2$	glycine	56-40-6	-3.21	-3.21	-3.41	-3.00	-3.08
$C_6H_{13}NO_2$	isoleucine	73-32-5	-1.76	-1.76	-1.59	-1.80	-1.88
$C_6H_{13}NO_2$	leucine	61-90-5	-1.74	-1.67	-1.59	-1.72	-1.92
$C_5H_{11}NO_2SM$	methionine	63-68-3	-2.00	-1.73	-2.41	-2.10	-2.27
$C_9H_{11}NO_2$	phenylalanine	63-91-2	-1.52	-1.56	-1.28	-1.36	-1.54
$C_5H_9NO_2$	proline	147-85-3	-2.43	-2.41	-2.15	-2.62	-3.11
$C_3H_7NO_3$	serine	56-45-1	-3.33	-2.74	-3.45	-2.73	-3.97
$C_4H_9NO_3$	threonine	72-19-5	-2.98	-2.43	-3.04	-2.43	-3.28
$C_{11}H_{12}N_2O_2$	tryptophan	73-22-3	-1.06	-1.57	-1.22	-1.01	-0.15
$C_9H_{11}NO_3$	tyrosine	60-18-4	-2.26	-2.23	-1.76	-1.95	-1.89
$C_5H_{11}NO_2$	valine	72-18-4	-2.26	-2.29	-2.08	-2.29	-2.39
$C_{14}H_{20}N_2O_4$	Val-Tyr	3061-91-4	-2.52	-1.43	-1.01	-2.44	-1.00
$C_{15}H_{22}N_2O_3$	Phe-Leu	3303-55-7	-1.17	-0.24	-0.49	-1.12	-0.10
$C_9H_{18}N_2O_3$	Ala-Ile	29727-65-9	-2.60	-1.65	-2.19	-2.61	-1.21
$C_{14}H_{18}N_2O_3$	Pro-Phe	13589-02-1	-2.07	-1.25	-0.96	-2.17	-1.01
$C_{11}H_{22}N_2O_3S$	Leu-Met	36077-39-1	-1.87	-1.51	-1.62	-1.86	-1.09
$C_{20}H_{21}N_3O_3$	Trp-Phe	6686-02-8	-0.47	-0.28	-0.12	-0.41	0.38
$C_{23}H_{26}N_4O_4$	Trp-Phe-Ala	126855-18-3	-1.00	-0.54	-0.53	-0.94	2.11
$C_{20}H_{23}N_3O_5$	Gly-Phe-Tyr	70421-71-5	-1.96	-1.51	-1.49	-2.27	1.10
$C_{15}H_{29}N_3O_5$	Thr-Val-Leu	66317-22-4	-1.97	-1.68	-1.74	-2.24	-1.06
$C_{15}H_{29}N_3O_5$	Ser-Leu-Ile	126855-27-4	-1.99	-1.46	-1.67	-2.05	-1.51
$C_{17}H_{31}N_3O_4$	Leu-Pro-Leu	126855-32-1	-1.56	0.36	-1.32	-1.46	-0.50
$C_{27}H_{31}N_5O_6$	Tyr-Pro-Gly-Trp	143313-24-0	-1.25	-0.97	-1.95	-1.81	-1.14
$C_{25}H_{40}N_4O_5S$	Val-Met-Phe-Ile	143303-25-7	-0.63	0.33	0.18	-0.18	1.90

Formula	Name	CAS	(a)	(b)	(c)	(d)	(e)
$C_{20}H_{30}N_4O_5$	Ile-Ala-Gly-Phe	14303-15-5	-1.78	-1.06	-1.73	-2.56	0.94
$C_{24}H_{38}N_4O_6$	Val-Phe-Leu-Thr	14313-22-8	-1.32	-0.58	-1.45	-1.83	1.10
$C_{20}H_{37}N_5O_6$	Gly-Ala-Ala-Leu-Leu	14303-37-1	-2.55	-1.28	-2.44	-3.43	0.30
$C_{25}H_{39}N_5O_6$	Gly-Leu-Leu-Gly-Phe	143303-41-7	-0.18	-0.17	-1.15	-2.25	1.13
$C_{32}H_{42}N_6O_6$	Trp-Leu-Phe-Ala-Ala	143303-48-4	-0.32	0.40	0.05	-1.08	2.20
$C_{23}H_{43}N_5O_7$	Ser-Leu-Ala-Ile-Val	143303-50-8	-1.94	-0.86	-1.58	-2.76	-1.69
$C_{27}H_{35}N_5O_7S$	Tyr-Gly-Gly-Phe-Met	—	-1.39	-2.43	-2.15	-2.70	0.28

(a) Recommended Value (Sangster, 1993).
(b) ClogP for Windows 1.0.0.
(c) KOWWIN 1.53.
(d) ACD/LogP 1.0.
(e) PrologD 2.0.

1990). Abraham and Leo (1995) discuss the particular case of acetylacetone in detail. Questions of tautomerism can arise in molecules such as 2-pyridinamine, 21 and 4-pyridinone, benzotriazoles, etc.

At present, fragment methods do not distinguish diastereomers (endo/exo, trans/gauche, cis/trans, etc.) although the partition coefficients are distinct. A particularly well illustrative study is that of Pleiss and Grunewald (1983) on substituted 1,4-methano-tetrahydronaphthalenes. Modification of fragment schemes to accommodate these effects of conformation were discussed by Rekker and Mannhold (1992) and Taylor (1990).

Some apparent fragment calculation anomalies are given in Table 6.13. Amiodarone (item 1) is a largish drug molecule (not excessively so), but all methods overestimate (there is no reliable experimental datum). Para-cyclophane (item 2) is also a puzzle. Earlier measured results indicated 3.70 (Hansch and Leo, 1979) but later values (Camilleri et al., 1988) are slightly higher. Evidently paracyclophane is not a close calculational equivalent to two p-xylene molecules. Questions of actual molecular shape may also be responsible for the different results for the cryptand 222 (item 7). The remaining items in Table 6.13 are discussed by Leo (1993, 1995b). The 2,6-disubstituted phenols (items 8 and 9) are illustrations of ortho effects (steric or otherwise) which have not been completely evaluated. In the condensed aromatic quinones (items 5 and 6), the hydrophilic nature of the carbonyls depend on their orientation relative to the rest of the molecule. There may be C1-O interaction in 4-chlorobutanol (item 4). Electronic effects act through a benzyl carbon atom (item 10). Dicofol (item 3) is a monohydroxy analogue of p,p'-DDT; the difference between measured and calculated values is striking. Apparently dicofol is more hydrophilic than calculation makes it out to be. This may be a case of hindered access by octanol to the hydroxyl (Leo, 1995b).

2 WHOLE MOLECULE METHODS

When a solute molecule enters solution, it does so as a whole and not as a series of fragments. The impeccable logic of this truism has convinced a number of investigators that molecular properties are a better route to calculating $\log K_{ow}$. Whole molecule methods, though of more recent origin than those based on fragments, have been shown to be successful and offer some insight in particular areas. A sample of predictions is summarized in Table 6.14.

Some of the results in this table are impressive, particularly the LSER method (Abraham et al., 1994). There are perhaps three basic decisions one must make when attempting these kinds of calculations: (1) which molecular properties to use; (2) how to calculate (or measure) these properties; and (3) which molecules to use in the calculations and regressions. Each step has

Table 6.7. Calculated log K_{ow} of end-protected-peptides

Formula	Name	CAS registry number	log K_{ow} expt. (a)	Hansch and Leo (b)	Meylan and Howard (c)	ACD (d)	Rekker (e)	Klopman (f)
$C_{20}H_{3}N_3O_4$	Ace-Tyr-Phe-NH$_2$	52329-50-7	0.54	0.23	0.95	0.59	0.43	1.82
$C_{11}H_{21}N_3O_4$	Ace-Thr-Val-NH$_2$	132765-89-0	−1.25	−1.56	−1.12	−1.58	−1.90	−1.21
$C_{11}H_{21}N_3O_3$	Ace-Ala-Leu-NH$_2$	78233-72-4	−0.54	−0.48	−0.58	−1.08	−0.80	−0.30
$C_{18}H_{24}N_4O_3$	Ace-Trp-Val-NH$_2$	132765-86-7	0.73	0.40	0.69	0.24	0.50	1.28
$C_{10}H_{19}N_3O_3$	Ace-Ser-Val-NH$_2$	132765-87-8	−1.53	−1.87	−1.54	−1.92	−2.42	−1.64
$C_{15}H_{28}N_4O_4$	Ace-Val-Gly-NH$_2$	132766-03-1	−0.45	−0.12	−0.50	−1.03	−0.83	−0.21
$C_{24}H_{27}N_5O_4$	Ace-Trp-Gly-Phe-NH$_2$	132766-16-6	0.99	0.32	0.66	0.42	0.06	1.64
$C_{18}H_{34}N_4O_5$	Ace-Leu-Thr-Leu-NH$_2$	132766-19-9	0.24	0.17	0.66	−0.11	−0.37	0.00
$C_{20}H_{30}N_4O_5$	Ace-Leu-Ser-Phe-NH$_2$	132766-18-8	0.23	−0.18	0.25	0.01	−0.78	0.15
$C_{20}H_{30}N_4O_5$	Ace-Ala-Tyr-Leu-NH$_2$	132766-20-2	−0.04	0.01	0.23	−0.36	−0.21	0.90

(a) Recommended Values from Sangster (1993).
(b) ClogP for Windows 1.0.0.
(c) KOWWIN 1.53.
(d) ACD/LogP 1.0.
(e) PrologP 5.1.
(f) KlogP contained in ToxAlert 1.2.

Table 6.8 Calculated log K_{ow} of pesticides

Formula	Name	CAS Registry Number	log K_{ow} expt. (a)	Hansch and Leo (b)	Meylan and Howard (c)	ACD (d)	Rekker (e)	Klopman (f)
Carbamates								
$C_{12}H_{15}NO_3$	carbofuran	1563-66-2	2.32	2.32	2.30	1/76	1.95	2.59
$C_{16}H_{20}N_4O$	diaziquone	57998-68-2	-0.02	-0.10	-0.59	-1.40	0.24	-0.25
$C_{14}H_{21}N_3O_3$	karbutilate	4849-32-5	1.66	1.47	1.93	1.66	1.21	2.05
$C_{11}H_{15}NOS$	methiocarb	2032-65-7	2.92	2.87	2.87	2.89	2.79	2.89
$C_7H_{13}N_3O_3S$	oxamyl	23135-22-0	-0.47	-0.47	-1.20	0.36	-0.27	1.63
$C_{11}H_{15}NO_3$	propoxur	114-26-1	1.52	1.65	1.90	1.60	2.26	1.96
DDT-type								
$C_{14}H_8Cl_4$	p,p'-DDE	72-55-9	6.96	6.94	6.00	6.51	7.33	6.91
$C_{14}H_9Cl_5$	p,p'-DDT	50-29-3	6.36	6.76	6.89	5.92	7.29	6.67
$C_{16}H_{15}Cl_3O_2$	methoxychlor	72-43-5	4.95	5.18	5.67	4.56	5.95	5.32
P-containing								
$C_9H_{11}Cl_3NO_3PS$	chlorpyrifos	2921-88-2	4.96	4.42	4.66	4.77	5.02	3.70
$C_8H_{16}NO_5P$	dicrotophos	141-66-2	0.00	0.05	-1.10	0.01	-0.64	0.30
$C_{10}H_{15}O_3PS_2$	fenthion	55-38-9	4.09	3.91	4.08	3.21	4.25	2.17
$C_{10}H_{14}NO_5PS$	parathion	56-38-2	3.83	3.47	3.73	3.84	3.91	1.71
$C_8H_{20}O_5P_2S_2$	sulfotep	3689-24-5	3.99	m.f.	3.98	2.24	3.11	0.51
$C_{12}H_{16}N_3O_3PS$	triazophos	24017-47-8	3.55	3.11	2.92	4.01	2.80	1.38
$C_8H_{18}NO_4PS_2$	vamidothion	2275-23-2	—	0.12	0.16	0.40	0.21	1.45

(Continued)

Simazine type

			(a)	(b)	(c)	(d)	(e)	(f)
$C_7H_{12}ClN_5$	simazine	122-34-9	2.18	2.09	2.40	0.69	2.24	2.36
$C_9H_5Cl_3N_4$	anilazine	101-05-3	3.00	2.93	3.64	3.07	3.93	3.58
$C_8H_{14}ClN_5$	atrazine	1912-24-9	2.61	2.40	2.82	1.03	2.76	2.66
$C_{10}H_{16}ClN_5O_2$	proglinazine ethyl	68228-18-2	—	2.15	2.36	1.13	2.86	2.59
$C_9H_{16}ClN_5$	terbuthylazine	5915-41-3	3.06	2.80	3.27	1.38	2.82	3.11

Urea type

			(a)	(b)	(c)	(d)	(e)	(f)
$C_{12}H_{13}ClN_2O$	buturon	3766-60-7	3.00	2.74	2.66	2.61	2.55	2.27
$C_{14}H_8ClF_2N_2O_2$	diflubenzuron	35367-38-5	3.88	3.95	3.59	2.49	2.78	3.23
$C_{10}H_{11}F_3N_2O$	fluometuron	2164-17-2	2.42	2.39	2.35	2.36	2.13	2.38
$C_{10}H_{17}N_3O_2$	isouron	55861-78-4	1.98	1.47	1.51	0.53	0.97	1.62
$C_9H_{10}Cl_2N_2O_2$	linuron	330-55-2	3.20	3.00	2.91	3.15	1.75	2.21

m.f.=missing fragment.
(a) Recommended Values from Sangster (1993).
(b) ClogP for Windows 1.0.0.
(c) KOWWIN 1.53.
(d) ACD/LogP 1.0.
(e) PrologP 5.1.
(f) KlogP contained in ToxAlert 1.2.

Table 6.9. Calculated log K_{ow} of nucleosides

Formula	Name	CAS registry number	log K_{ow} expt. (a)	Hansch and Leo (b)	Meylan and Howard (c)	ACD (d)	Ghose and Crippen (e)	Rekker (f)	Klopman (g)
Uridines									
$C_9H_{12}N_2O_6$	parent	58-96-8	−1.98	−2.83	−1.86	−1.78	−1.59	−2.59	−2.17
$C_9H_2NO_5$	dUrd	951-78-0	−1.62	−2.44	−1.19	−1.47	−0.98	−2.07	−1.37
$C_9H_{12}N_2O_4$	ddUrd	5983-09-5	−0.89	−1.02	−1.12	−1.18	−0.54	−0.98	−0.56
$C_9H_{10}N_2O_4$	dde	5974-93-6	−1.07	−0.98	−1.34	−1.26	−0.32	−1.34	−0.48
$C_9H_{11}FN_2O_4$	Fdd	41107-56-6	−0.49	−1.18	−1.26	−1.15	−0.37	−1.04	−0.31
Cytidines									
$C_9H_{13}N_3O_5$	parent	65-46-3	−2.51	−3.08	−2.46	−2.10	−1.42	−4.34	−3.25
$C_9H_{13}N_3O_4$	dCyd	951-77-9	−1.77	−2.73	−1.79	−1.79	−0.81	−3.82	−2.44
$C_9H_{13}N_3O_3$	ddCyd	7481-89-2	−1.30	−1.32	−1.73	−1.50	−0.37	−2.72	−1.64
$C_9H_{11}N_3O_3$	dde	7481-88-1	−1.57	−1.27	−1.94	−1.59	−0.14	−3.09	−1.56
$C_9H_{12}FN_3O_3$	Fdd	51246-79-8	−0.92	−1.47	−1.86	−1.51	−0.20	−2.79	−1.38
Adenosines									
$C_{10}H_{13}N_5O_4$	parent	58-61-7	−1.23	−2.88	−1.38	−1.26	−1.24	−2.22	−2.52
$C_{10}H_{13}N_5O_3$	dAdo	958-09-8	−0.55	−2.29	−0.71	−0.53	−0.63	−1.70	−1.72
$C_{10}H_{13}N_5O_2$	ddAdo	4097-22-7	−0.22	−0.88	−0.65	−0.49	−0.19	−0.60	−0.91
$C_{10}H_{11}N_5O_2$	ddeAdo	7057-48-9	−0.36	−0.83	−0.86	−0.56	0.04	−0.97	−0.83
$C_{10}H_{12}FN_5O_2$	ddeF	87418-35-7	0.08	−1.04	−0.78	−0.49	−0.02	−0.67	−0.66

(Continued)

		CAS	(a)	(b)	(c)	(d)	(e)	(f)	(g)
Guanosines									
$C_{10}H_{13}N_5O_5$	parent	118-00-3	−1.89	−3.85	−1.71	−1.89	−1.63	−2.41	−3.03
$C_{10}H_{13}N_5O_4$	dUrd	961-07-9	−1.30	−3.27	−1.04	−1.28	−1.01	−1.89	−2.2
$C_{10}H_{13}N_5O_3$	ddUrd	85326-06-3	−1.01	−1.85	−0.97	−0.73	−0.58	−0.80	−1.42
$C_{10}H_{11}N_5O_3$	ddeUrd	53766-80-6	−1.21	−1.81	−1.19	−0.80	−0.35	−1.16	−1.33
Thymidines									
$C_{10}H_{14}N_2O_5$	parent	50-89-5	−1.17	−1.94	−0.64	−1.18	−0.82	−1.55	−0.98
$C_{10}H_{14}N_2O_4$	3'-deoxy	3416-05-5	−0.63	−0.52	−0.58	−0.82	−0.39	−0.46	−0.17
$C_{10}H_{12}N_2O_4$	3'-deoxy-2',3'-didehydro-	3056-17-5	−0.81	−0.48	−0.79	−0.90	−0.16	−0.82	−0.09
$C_{10}H_{13}FNO_4$	3'-deoxy-3'-F-	25526-93-6	−0.28	−0.68	−0.71	−0.82	−0.22	−0.52	0.08

Ado = Adenosine
Cyd = Cytidine
Guo = Guanosine
Thd = Thymidine
Urd = Uridine

d = 2'-deoxy
dd = 2',3'-dideoxy-
dde = 2',3'-dideoxy- 2',3'-didehydro-

F = 3'-fluoro-

(a) Recommended Values from Sangster (1993).
(b) ClogP for Windows 1.0.0.
(c) KOWWIN 1.53.
(d) ACD/LogP 1.0.
(e) Viswanadhan et al. (1993)
(f) PrologP 5.1.
(g) KlogP contained in ToxAlert 1.2.

Table 6.10 Calculated log K_{ow} of polychlorinated aromatics

Formula	Substitution	CAS registry number	log K_{ow} expt. (a)	Hansch and Leo (b)	Meylan and Howard (c)	ACD (d)	Rekker (e)	Klopman (f)
PCBs								
$C_{12}H_{10}$	none	92-52-4	3.98	4.03	3.76	3.98	3.97	4.09
$C_{12}H_9Cl$	4-	2051-62-9	4.61	4.74	4.40	4.55	4.71	4.02
$C_{12}H_8Cl_2$	2,2'-	13029-08-8	4.73	4.96	5.05	4.93	5.45	4.68
$C_{12}H_8Cl_2$	4,4'-	2050-68-2	5.58	5.46	5.05	5.12	5.45	4.68
$C_{12}H_7Cl_3$	2,4,5-	15862-07-4	5.81	5.80	5.69	5.40	6.19	5.34
$C_{12}H_7Cl_3$	2,2',5-	37680-65-2	5.60	5.67	5.69	5.43	6.19	5.34
$C_{12}H_6Cl_4$	2,2',5,5'-	35693-99-3	6.09	6.38	6.34	5.92	6.93	5.99
$C_{12}H_6Cl_4$	2,3,4,5-	33284-53-6	6.41	6.27	6.34	5.73	6.93	5.99
$C_{12}H_5Cl_5$	2,2',4,5,5'-	37680-73-2	6.44	7.00	6.98	6.38	7.67	6.65
$C_{12}H_5Cl_5$	2,3,4,5,6-	18259-05-7	6.52	6.62	6.98	6.04	7.67	6.65
$C_{12}H_4Cl_6$	2,2',4,4',5,5'-	35065-27-1	6.80	7.57	7.62	6.82	8.41	7.30
$C_{12}H_4Cl_6$	3,3',4,4',5,5'-	32774-16-6	7.55	7.83	7.62	6.90	8.41	7.30
$C_{12}H_3Cl_7$	2,2',3,3',4,4',6-	52663-71-5	6.99	8.25	8.26	7.16	9.15	7.96
$C_{12}H_2Cl_8$	2,2',3,3',5,5',6,6'-	2136-99-4	7.15	8.25	8.91	7.41	9.89	8.61
$C_{12}HCl_9$	2,2',3,3',4,5,5',6,6'-	52663-77-1	8.16	8.73	9.56	7.76	9.89	9.27
$C_{12}Cl_{10}$	decachloro	2051-24-3	8.26	9.20	10.2	8.10	11.4	9.92

(Continued)

PCDDs

Formula	Substitution	CAS	(a)	(b)	(c)	(d)	(e)	(f)
$C_{12}H_8O_2$	none	262-12-4	4.37	4.62	4.34	3.34	4.30	3.30
$C_{12}H_7ClO_2$	1-	39227-53-7	5.05	5.39	4.97	4.03	5.04	3.96
$C_{12}H_6Cl_2O_2$	2,7-	33857-26-0	6.38	6.12	5.63	5.27	5.78	4.61
$C_{12}H_5Cl_3O_2$	1,2,4-	39227-58-2	7.47	6.71	6.28	5.50	6.52	5.27
$C_{12}H_4Cl_4O_2$	2,3,7,8-	1746-01-6	6.42	7.31	6.92	6.68	7.26	5.92
$C_{12}H_3Cl_5O_2$	1,2,3,7,8-	40321-76-4	6.64	7.90	7.56	7.25	8.00	6.58

PCDFs

Formula	Substitution	CAS	(a)	(b)	(c)	(d)	(e)	(f)
$C_{12}H_8O$	none	132-64-9	4.12	4.09	4.05	4.12	4.39	3.32
$C_{12}H_6Cl_2O$	2,8-	5409-83-6	5.65	5.51	5.34	5.31	5.87	4.63
$C_{12}H_4Cl_4O$	1,2,7,8-	58802-20-3	6.23	6.70	6.63	6.25	7.35	5.94
$C_{12}H_3Cl_5O$	1,2,4,6,8-	69698-20-3	6.34	7.53	7.27	6.92	8.09	6.59
$C_{12}H_2Cl_6O$	1,2,3,6,8,9-	75198-38-8	6.82	8.00	7.92	7.27	8.83	7.25
$C_{12}HCl_7O$	1,2,3,4,6,7,8-	67562-39-4	7.92	8.48	8.56	7.62	9.57	7.90

PCBs = polychlorinated biphenyls
PCDDs = polychlorinated dibenzodioxins
PCDFs = polychlorinated debenzofurans
(a) Recommended Values from Sangster (1993).
(b) ClogP for Windows 1.0.0.
(c) KOWWIN 1.53.
(d) ACD/LogP 1.0.
(e) PrologP 5.1.
(f) KlogP contained in ToxAlert 1.2.

Table 6.11. Calculated log K_{ow} of organic dyes

Formula	Name	CAS registry number	log K_{ow} expt. (a)	Hansch and Leo (b)	Meylan and Howard (c)	ACD (d)	Rekker (e)	Klopman (f)
$C_{23}H_{25}BrN_6O_{10}$	Disp. Blue 79	3956-55-6	3,6	4.28	5.04	4.26	5.25	5.95
$C_{17}H_{19}ClN_4O_4$	Disp. Red 5	3769-57-1	3,4,4.3	3.71	4.33	4.04	4.84	5.14
$C_{15}H_{15}N_3O_2$	Disp. Yellow 5	2823-40-8	2.9	3.68	3.98	2.92	3.93	4.70
$C_{21}H_{20}BrN_7O_6$	N1	68877-63-4	2.5,5,4,7.3	4.12	5.35	4.47	5.12	6.08
$C_{15}H_{12}ClN_5O_4$	N2	70528-90-4	3.4,4.5	3.04	3.57	4.32,2.86	2.52	4.11
$C_{23}H_{24}N_6O_4$	N5	42828-64-9	4.0,5,5.7,8	5.13	6.79	6.79	6.36	6.47
$C_{17}H_{17}Cl_2N_5O_4$	N7	71617-28-2	4.0,5,4,5.8	3.63	4.38	4.55	4.69	5.83
$C_{16}H_{16}ClN_5O_3$	N9	6657-33-6	3.9,4.0,4.9	3.48	4.38	3.88	4.74	5.43
$C_{23}H_{25}BrN_6O_{10}$	Ra	3956-55-6	3.6	4.28	5.04	4.26	5.25	5.95
$C_{12}H_{10}N_2O$	Solvent Yellow 7	1689-82-3	4.0,5,5,7.8	5.13	6.79	6.79	6.36	6.47
Anthraquinone type								
$C_{14}H_8O_4$	alizarine	72-48-0	2.3	1.99	3.16	3.33	1.49	3.13
$C_{14}H_6Cl_2N_2O_4$	Disp. A 18	66121-41-3	—	3.15	5.21	3.50	2.72	3.80
$C_{14}H_{12}N_4O_2$	Disp. Blue 1	2475-45-8	−0.96	−1.65	2.98	1.71	−1.61	0.25
$C_{17}H_{16}N_2O_3$	Disp. Blue 3	2475-46-9	1.9	1.51	3.28	2.80	1.14	3.46
$C_{16}H_{14}N_2O_2$	Disp. Blue 14	2475-44-7	2.7	2.27	4.26	3.74	1.71	3.48
$C_{15}H_{11}NO_2$	Disp. Orange	82-28-0	2.8	2.29	4.07	2.97	2.02	2.34
$C_{20}H_{14}N_2O_3$	Disp. Red 60	41603-04-7	6.2,7.2	3.06	4.38	4.08	2.61	2.74
$C_{14}H_9N_3O_4$	Disp. Violet 8	82-33-7		0.69	2.98	1.67	0.20	1.43
$C_{14}H_8O_5$	Purpurin	81-54-9	1.8	1.37	3.46	3.92	0.96	3.77
$C_{20}H_{17}N_3O_4$	N3	3176-88-3	4.0,5,8,7.5	1.78	4.15	5.64	−0.38	1.59
$C_{12}H_{13}NO_3$	Disp. Red 3	4465-58-1	2.1	1.86	3.10	2.35	1.55	2.29
Other								
$C_{12}H_9N_3O_5$	Disp. Yellow 1	114-15-3	3.9	3.28	2.67	2.87	2.60	3.14
$C_{16}H_{10}N_2O_2$	Indigo	482-89-3	3.72R	2.70	3.11	3.19	2.80	2.68
$C_{28}H_{14}N_2O_4$	Vat Blue 4	81-77-6	—	4.61	7.73	7.46	4.25	4.78

(a) Experimental data from Sangster (1993). R indicates a Recommended Value.
(b) ClogP for Windows 1.0.0.
(c) KOWWIN 1.53.
(d) ACD/LogP 1.0. Where two data are given, these are for separate tautomers.
(e) PrologP 5.1.

Table 6.12 Calculated log K_{ow} of radiopaque compounds

Formula	Name	CAS registry Number	log K_{ow} expt. (a)	Hansch and Leo (b)	Meylan and Howard (c)	ACD (d)	Rekker (e)	Klopman (f)
$C_{17}H_{22}I_3N_3O_7$	—	74411–74–8	−1.86	−4.09	−0.71	−3.96	0.39	0.87
$C_{17}H_{22}I_3N_3O_7$	—	88116–55–6	−2.06	−4.49	−0.79	−3.55	0.39	1.13
$C_{17}H_{22}I_3N_3O_8$	iopamidol	60166–93–0	−2.42	−6.81	−1.38	−2.43	0.24	0.38
$C_{17}H_{22}I_3N_3O_8$	P569	—	−2.64	−5.13	−2.51	−5.33	−1.22	−0.72
$C_{17}H_{22}I_3N_3O_9$	—	31122–84–6	−2.17	−2.17	−2.69	−3.51	−0.28	0.24
$C_{18}H_{24}I_3N_3O_7$	—	66108–93–8	−2.47	−4.92	−1.76	−4.49	−0.18	0.37
$C_{18}H_{24}I_3N_3O_8$	—	88116–58–9	−2.33	−2.30	−0.35	−3.81	−0.18	0.63
$C_{18}H_{24}I_3N_3O_8$	iopromide	73334–07–3	−2.05	−4.61	−2.49	−3.24	−0.07	−/94
$C_{18}H_{24}I_3N_3O_9$	ioversol	87771–40–2	−2.98	−5.24	−2.32	−5.29	−0.99	−0.57
$C_{18}H_{24}I_3N_3O_9$	P530	119425–66–0	−2.55	−2.55	−2.12	−5.12	−1.19	−0.47
$C_{19}H_{26}I_3N_3O_9$	iohexol	66108–95–0	−3.05	−5.74	−2.81	−5.43	−0.79	−0.14

(a) Recommended Values from Sangster (1993).
(b) ClogP for Windows 1.0.0.
(c) KOWWIN 1.53.
(d) ACD/LogP 1.0.
(e) PrologP 5.1.
(f) KlogP contained in ToxAlert 1.2.

Table 6.13 Apparent anomalies in log K_{ow} calculation

No.	Formula	Name	CAS registry Number	log K_{ow} expt. (a)	Hansch and Leo (b)	Meylan and Howard (c)	ACD (d)	Rekker (e)	Klopman (f)
1.	$C_{25}H_{29}I_2NO_3$	amiodarone	1951–25	–	8.75	8.81	8.58	9.84	8.55
2.	$C_{16}H_{16}$	paracyclophane	1633–22–3	4.47	5.19	5.72	5.91	5.39	5.04
3.	$C_{14}H_9Cl_5O$	dicofol	115–32–2	4.28	6.06	5.81	5.74	6.02	6.01
4.	C_4H_9ClO	4-chlorobutanol	928–51–8	0.85	0.54	1.10	0.85	0.65	1.09
5.	$C_{14}H_8O_2$	anthraquinone	84–65–1	3.39	2.62	3.34	2.44	2.54	2.67
6.	$C_{14}H_8O_2$	phenanthraquinone	84–11–7	2.52	3.25	3.56	3.13	2.90	2.67
7.	$C_{18}H_{36}N_2O_6$	222	23978–09–8	–0.10	–0.79	–2.14	–1.34	0.74	2.68
8.	$C_{14}H_{22}O$	2, 6-di(t-butyl)phenol	128–39–2	4.36	5.40	4.48	4.86	5.65	4.59
9.	$C_{14}H_{22}O$	2, 6-dii(s-butyl)phenol	5510–99–6	4.38	5.39	4.48	5.22	5.65	4.52
10.	$C_{18}H_{17}N_3O_5$	α-cyanophenyl-amphenicol	49648–48–8	1.47	0.66	1.73	1.50	0.47	1.62

(a) Recommended Values from Sangster (1993).
(b) ClogP for Windows 1.0.0.
(c) KOWWIN 1.53.
(d) ACD/LogP 1.0.
(e) PrologP 5.1.
(f) KlogP contained in ToxAlert 1.2.

Table 6.14 Calculation of log K_{ow} by molecular properties

Compounds	Parameters	n	r	s	F	log K_{ow} range	Reference
Connectivity indices							
N bridge compounds	$^1\chi$, $^4\chi_p^v$	56	0.962	0.17	336	0.20, 2.44	Szasz et al. (1983)
PCBs, etc.	$^1\chi^v$	64	0.982	0.29	813	2.13, 8.20	Doucette and Andren (1988)
non-H bonders	7 parameters	103	0.912	0.47	–	–	Niemi et al. (1992)
PCBs	CRI	34	0.997	0.12	1538	3.89, 8.28	Saçan and Inel (1995)
Surface area							
Various	2 parameters	138	0.995	0.13	7285	−1.93, 4.88	Iwase et al. (1985)
Various	5 parameters	102	0.992	0.11	1192	–	Chastrette et al. (1990)
PCBs	TSA	46	0.959	0.32	–	3.76, 8.20	Hawker and Connell (1988)
Thermodynamic quantities							
PCBs	$\Delta_{tr}G$(water → octanol)	15	0.983	0.15	350	3.90, 6.50	Miertuš and Jakuš (1990)
Composite							
Phosphorothionates	$\Delta_{tr}G$(gas → water), CSA	12	0.970	0.23	–	2.71, 5.51	Schüürmann et al. (1995a)
Various	4 parameters	70	0.985	0.28	–	−1.38, 5.18	Brinck et al. (1993)
Various	$\Delta_{tr}G$(gas → water), CSA	17	0.954	0.24	–	0.90, 3.89	Schüürmann et al. (1995b)
Various	6 parameters	244	0.976	0.42	787	−3.30, 5.80	Kasai et al. (1988)
Pesticides	3 parameters	67	0.977	0.43	442	−3.60, 6.31	Katagi et al. (1995)
Various	LSER	613	0.997	0.12	23162	−1.51, 8.00	Abraham et al. (1994)

CRI = characteristic root index; CSA = contact surface area; TSA = total surface area; χ = Kier connectivity index attendant uncertainties, and it is as yet unclear to what extent these decisions are interdependent.

3 FRAGMENTS OR MOLECULAR PROPERTIES?

Fragment methods are sometimes criticized for being 'merely' empirical. This makes as much sense as criticizing molecular orbital (MO) methods for being 'merely' theoretical. Molecular properties may be the wave of the future, but at the present time the fragment methods are applicable to a wide variety of organic compounds, and the algorithms are compact enough to be suited for PC use. None of the molecular properties methods have been applied to drug molecules of any complexity, and PC applications are impractical.

Molecular properties methods, however, may prove their strength in being able to distinguish the same molecule in different conformations. This applies not only to small molecules (Kantola et al., 1991; Kasai et al., 1988) but also to larger flexible species such as crown ethers and similar macrocyclics (Georgiu et al., 1982; Pigot et al., 1992; Stolwijk et al., 1989). Fragment methods can of course be modified to take conformation and flexibility into account (Richards et al., 1991, 1992). It might, however, be easier for MO rather than fragment methods to do this. In judging the success of different methodologies, several aspects are pertinent: accuracy of prediction, understanding of underlying physicochemical effects and quality of known experimental data. The relative importance one assigns to these aspects can lead to the espousal of distinct but complementary viewpoints (Kamlet et al., 1987).

REFERENCES

Abraham, M. H., Chadha, H. S., Whiting, G. S. and Mitchell, R. C. (1994) *J. Pharm. Sci.* **83**, 1085–100.
Abraham, M. H. and Leo, A. J. (1995) *J. Chem. Soc. Perkin Trans.* 2, 1839–42.
Brinck, T., Murray, J. S. and Politzer, P. (1993) *J. Org. Chem.* **58**, 7070–3.
Budavari, S., ed. (1996) *The Merck Index*, Merck and Co., Whitehouse Station.
Camilleri, P., Watts, S. A. and Boraston, J. A. (1988) *J. Chem. Soc. Perkin Trans.* 2, 1699–1707.
Chastrette, M., Tiyal, F. and Peyraud, J.-F. (1990) *C. R. Acad. Sci. Paris, Ser. II* **311**, 1057–60.
Doucette, W. J. and Andren, A. W. (1988) *Chemosphere* **17**, 345–59.
Dross, K. P., Mannhold, R. and Rekker, R. F. (1992) *Quant. Struct.-Act. Relat.* **11**, 36–44.
Elks, J. and Ganellin, C. R. (1990) *Dictionary of Drugs*, 2 vols., Chapman and Hall, London.
Georgiu, P., Richardson, C. H., Simmons, K., Truter, M. R. and Wingfield, J. N. (1982) *Inorg. Chim. Acta* **66**. 1–6.
Hansch, C. (gen. ed) (1990) *Comprehensive Medicinal Chemistry*, 6 vols., Pergamon Press, New York.

Hansch, C. and Leo, A. J. (1979) *Substituent Constants for Correlation Analysis in Chemistry and Biology*, Wiley Interscience, New York.

Hansch, C. and Leo, A. J. (1995) *Exploring QSAR: fundamentals and applications in chemistry and biology*, American Chemical Society, Washington.

Hartley, D. and Kidd, H (eds) (1987) *The Agrochemicals Handbook*, 2nd edition, Royal Society of Chemistry, Nottingham.

Hawker, D. W. and Connell, D. W. (1988) *Environ. Sci. Technol.* **22**, 382–7.

Howard, P. H. and Neal, M. (1992) *Dictionary of Chemical Names and Synonyms*, Lewis Publishers, Chelsea.

Iwase, K., Komatsu, K., Hirono, S., Nakagawa, S. and Moriguchi, I. (1985) *Chem. Pharm. Bull.* **33**, 2114–21.

Kamlet, M. J., Doherty, R. M., Famini, G. R. and Taft, R. W. (1987) *Acta Chem. Scand. B* **41**, 589–98.

Kantola, A., Villar, H. O. and Loew, G. H. (1991) *J. Comput. Chem.* 12, 681–9.

Kasai, K., Umeyama, H. and Tomonaga, A. (1988) *Bull. Chem. Soc. Jpn.* 61, 2701–6.

Katagi, T., Miyakado, M., Takayama, C. and Tanaka, S. (1995) *ACS Symp. Ser.* **606**, 48–61.

Kühne, R., Rothenbacher, C., Herth, P. and Schüürmann, G. (1994) in C. Jochum, (ed), *Software Development in Chemistry 8*, Gesellschaft Deutscher Chemiker, Frankfurt/Main, pp. 207–24.

Leo, A. J. (1990) in Hansch (1990), vol. 4, pp. 295–319.

Leo, A. J. (1993) *Chem Rev.* **93**, 1281–1306.

Leo, A. J. (1995a) *Chem. Pharm. Bull.* **43**, 512–13.

Leo, A. J. (1995b) *ACS Symp. Ser.* **606**, 62–74.

Mannhold, R., Dross, K. P. and Rekker, R. F. (1990) *Quant. Struct.-Act. Relat.* **9**, 21–8.

Mannhold, R., Rekker, R. F. Sonntag, C., ter Laak, A. M., Dross, K. and Polymeropoulos, E. E. (1995) *J. Pharm. Sci.* **84**, 1410–19.

Mayer, J. M., van de Waterbeemd, H. and Testa, B. (1982) *Eur. J. Med. Chem.-Chim. Ther.* **17**, 17–25.

Miertuš, S. and Jakuš, V. (1990) *Chem Pap.* **44**, 793–804.

Morigulchi, I., Hirono, S., Nakagome, I. and Hirano, H. (1994) *Chem. Pharm. Bull.* **42**, 976–8.

Negwer, M. (1987) *Organic-Chemical Drugs and Their Synonyms*, 6th edition, 3 vols., VCH Publishers, New York., VCH Publishers, New York.

Niemi, G. J., Basak, S. C., Veith, G. D. and Grunwald, G. (1992) *Environ. Toxicol. Chem.* **11**, 893–900.

Nys, G. G. and Rekker, R. F. (1974) *Eur. J. Med. Chem.-Chim. Ther.* **9**, 361–75.

Pigot, T., Duriez, M.-C., Cazaux, L. and Tisnes, P. (1992) *Tetrahedron* **48**, 4359–68.

Pleiss, M. A. and Grunwald, G. L. (1983) *J. Med. Chem.* 26, 1760–4.

Rekker, R. F. and De Kort, H. M. (1979) *Eur. J. Med. Chem.-Chim. Ther.* **14**, 479–88.

Rekker, R. F., ter Laak, A. M. and Mannhold, R. (1993) *Quant. Struct.-Act. Relat.* **12**, 152–7.

Rekker, R. F. and Mannhold, R. (1992) *Calculation of Drug Lipophilicity*, VCH Verlagsgesellschaft, Weinheim.

Richards, N. G. J., Williams, P. B. and Tute, M. S. (1991) *Int. J. Quantum Chem., Quantum Biol. Symp.*, **18**, 229–316.

Richards, N. G. J., Williams, P. B. and Tute, M. S. (1992) *Int. J. Quantum. Chem.* **44**, 219–33.

Saçan, M. T. and İnel, Y. (1995) *Chemosphere* **30**, 39–50.

Sangster, J. (1993) *LOGKOW—a databank of evaluated octanol–water parition coefficients*, Sangster Research Laboratories, Montreal.

Schüürmann, G. (1995a) *Fresenius Environ. Bull.* **4**, 238–43.

Schüürmann, G. (1995b) *Environ. Toxicol. Chem.* **14**, 2067–76.

Schüürmann, G., Kühne, R., Ebert, R.-U. and Kleint, F. (1995) *Fresenius Environ. Bull.* **4**, 13–18.

Society of Dyers and Colourists. (1971) *Colour Index*, 3rd edition, 5 vols., Society of Dyers and Colourists, Bradford. Revision volumes 6, 7 and 8 (1975, 1982 and 1987).

Stolwijk, T. B., Vos, L. C., Sudholter, E. J. R. and Reinhoudt, D. N. (1989) *Rec. Trav. Chem. Pays-Bas* **108**, 103–8.

Szász, G., Novák-Hankó, K., Kier, L. B., Hermecz, I. and Kökösi, J. (1983) *Acta Pharm. Hung.* **53**, 195–202.

Taylor, P. J. (1990) in Hansch (1990), vol. 4, pp. 241–94.

Viswanadhan, V. N., Reddy, M. R., Bacquet, R. J. and Erion, M. D. (1993) *J. Comput. Chem.* **14**, 1019–26.

Weininger, S. (1988) *J. Chem. Inf. Comput. Sci,* **28**, 31–6.

Worthing, C. R. (ed) (1987) *The Pesticide Manual*, 8th edition, British Crop Protection Council, Thornton Heath.

APPENDIX

TABLE OF RECOMMENDED LOG K_{OW} DATA

The following data are excerpted from LOGKOW, the author's Databank of evaluated partition coefficients. The complete Databank is available on diskette from the author, and also as an on-line version. Addresses for inquiries are given at the end of the Table.

Formula	CAS Registry Number	Name	log K_{ow}
CCl_2F_2	75-71-8	Dichlorodifluoromethane	2.16
CCl_3F	75-69-4	Flurotrichloromethane	2.53
CCl_4	56-23-5	Carbon tetrachloride	2.83
CF_4	75-73-0	Tetrafluoromethane	1.18
CO_2	124-38-9	Carbon dioxide	0.83
CS2	75-15-0	Carbon disulphide	1.94
$CHBrCl_2$	75-27-4	Bromodichloromethane	2.00
$CHBr_2Cl$	124-48-1	Chlorodibromomethane	2.16
$CHBr_3$	75-25-2	Bromoform	2.67
$CHCl_3$	67-66-3	Chloroform	1.97
CH_2Cl_2	75-09-2	Methylenechloride	1.25
CH_2O	50-00-0	Formaldehyde	0.35
CH_2O_2	64-18-6	Formic acid	−0.54
CH_3Br	74-83-9	Bromomethane	1.19
CH_3Cl	74-87-3	Methyl chloride	0.91
CH_3NO	75-12-7	Formamide	−1.51
CH_3NO_2	75-52-5	Nitromethane	−0.33
CH_4	74-82-8	Methane	1.09
CH_4N_2O	57-13-6	Urea	−2.11
CH_4O	67-56-1	Methanol	−0.74
CH_5N	74-89-5	Methylamine	−0.57
$C_2Cl_2F_4$	76-14-2	1,2-dichlorotetrafluoroethane	2.82
C_2Cl_4	127-18-4	Tetrachloroethylene	3.40
C_2Cl_6	67-72-1	Hexachloroethane	4.14

Continued

Formula	CAS Registry Number	Name	log K_{ow}
C_2HCl_3	79-01-6	Trichloroethylene	2.42
C_2H_2	74-86-2	Acetylene	0.37
$C_2H_2Cl_2$	75-35-4	1,1-Dichloroethylene	2.13
$C_2H_2Cl_2$	156-60-5	trans-1,2,-Dichloroethylene	2.09
$C_2H_2Cl_4$	79-34-5	1,1,2,2,-Tetrachloroethane	2.39
C_2H_3Cl	75-01-4	Vinyl chloride	1.52
$C_2H_3Cl_3$	71-55-6	1,1,1-Trichloroethane	2.49
$C_2H_3Cl_3$	79-00-5	1,1,2-Trichloroethane	1.89
C_2H_3N	75-05-8	Acetonitrile	−0.34
$C_2H_4Br_2$	106-93-4	Ethylenedibromide	1.96
$C_2H_4Cl_2$	75-34-3	1,1-Dichloroethane	1.79
$C_2H_4Cl_2$	107-06-02	1,2-Dichloroethane	1.48
C_2H_4O	75-21-8	Ethylene oxide	−0.30
C_2H_4O	75-07-0	Acetaldehyde	0.45
$C_2H_4O_2$	64-19-7	Acetic acid	−0.17
$C_2H_4O_2$	107-31-3	Methyl formate	0.03
C_2H_5NO	60-35-5	Acetamide	−1.26
$C_2H_5NO_2$	56-40-6	Glycine	−3.21
$C_2H_5NO_2$	79-24-3	Nitroethane	0.18
C_2H_6	74-89-0	Ethane	1.81
$C_2H_6N_2O$	62-75-9	N-Nitrosodimethyl amine	−0.57
C_2H_6O	64-17-5	Ethanol	−0.30
C_2H_6O	115-10-6	Dimethyl ether	0.10
C_2H_6OS	67-68-5	Dimethylsulphoxide	−1.35
$C_2H_6O_2$	107-21-1	Ethyleneglycol	−1.36
$C_2H_6S_2$	624-92-0	Dimethyl disulphide	1.77
C_2H_7N	124-40-3	Dimethyl amine	−0.38
C_2H_7N	75-04-7	Ethylamine	−0.13
C_2H_7NO	141-43-5	Ethanolamine	−1.31
$C_2H_8N_2$	107-15-3	Ethylenediamine	−2.04
C_3H_3N	107-13-1	Acrylonitrile	0.25
C_3H_4	74-99-7	Propyne	0.94
$C_3H_4N_2$	288-32-4	Imidazole	−0.08
$C_3H_4N_2$	288-13-1	Pyrazole	0.26
C_3H_4O	107-02-8	Acrolein	−0.01
$C_3H_4O_2$	79-10-7	Acrylic acid	0.35
C_3H_5N	107-12-0	Ethylcyanide	0.16
C_3H_6	75-19-4	Cyclopropane	1.72
$C_3H_6Cl_2$	78-87-5	1,2-Dichloropropane	2.02
$C_3H_6Cl_2$	142-28-9	1,3-Dichloropropane	2.00
$C_3H_6N_2S$	96-45-7	Ethylene thiourea	−0.66
C_3H_6O	67-64-1	Acetone	−0.24
C_3H_6O	123-30-6	Propanal	0.59
$C_3H_6O_2$	79-20-9	Methyl acetate	0.18
$C_3H_6O_2$	646-06-0	1,3-Dioxolane	−0.37
$C_3H_6O_2$	79-09-4	Propanoic acid	0.33
$C_3H_6O_2$	109-94-4	Ethyl formate	0.27

Formula	CAS Registry Number	Name	log K_{ow}
$C_3H_6O_3$	110-88-3	1,3,5-Trioxane	−0.43
C_3H_7NO	68-12-2	N,N-Dimethylformamide	−1.01
$C_3H_7NO_2$	56-41-7	L-Alanine	−2.85
$C_3H_7NO_2$	79-46-9	2-Nitropropane	0.80
$C_3H_7NO_3$	56-45-1	L-Serine	−3.33
C_3H_8	74-98-6	Propane	2.36
C_3H_8O	71-23-8	1-Propanol	0.25
C_3H_8O	67-63-0	2-Propanol	0.05
$C_3H_8O_2$	109-87-5	Methylal	0.18
$C_3H_8O_2$	109-86-4	2-Methoxyethanol	−0.77
$C_3H_8O_2$	57-55-6	Propylene glycol	−0.92
$C_3H_8O_3$	56-81-5	Glycerol	−1.76
C_3H_9N	107-10-8	Propylamine	0.48
C_3H_9N	75-50-3	Trimethylamine	0.16
C_4Cl_6	87-68-3	Hexachlorobutadiene	4.78
C_4H_4O	110-00-9	Furan	1.34
C_4H_4S	110-02-1	Thiofuran	1.81
C_4H_6	106-99-0	1,3-Butadiene	1.99
C_4H_6	107-00-6	1-Butyne	1.46
$C_4H_7Cl_2O_4P$	62-73-7	Dichlorvos	1.43
C_4H_8	106-98-9	1-Butene	2.40
$C_4H_8Cl_2O$	111-44-4	Bis(2-chloroethyl)ether	1.29
C_4H_8O	123-72-8	Butanal	0.88
C_4H_8O	78-93-3	Methylethyl ketone	0.29
C_4H_8O	109-99-9	Tetrahydrofuran	0.46
$C_4H_8O_2$	123-91-1	1,4-Dioxane	−0.27
$C_4H_8O_2$	107-92-6	Butanoic acid	0.79
$C_4H_8O_2S$	126-33-0	Sulpholane	−0.77
C_4H_9NO	127-19-5	N,N-Dimethylacetamide	−0.77
$C_4H_9NO_3$	72-19-5	L-Threonine	−2.98
C_4H_{10}	106-97-8	Butane	2.89
C_4H_{10}	75-28-5	Isobutane	2.76
$C_4H_{10}N_2O$	55-18-5	N-Nitrosodiethyl amine	0.48
$C_4H_{10}O$	71-36-3	1-Butanol	0.84
$C_4H_{10}O$	78-92-2	2-Butanol	0.65
$C_4H_{10}O$	78-83-1	Isobutanol	0.76
$C_4H_{10}O$	75-65-0	tert-Butanol	0.35
$C_4H_{10}O$	60-29-7	Diethyl ether	0.89
$C_4H_{10}O_2$	110-80-5	2-Ethoxyethanol	−0.28
$C_4H_{11}N$	109-73-9	Butylamine	0.86
$C_4H_{11}N$	75-64-9	tert-Butylamine	0.40
$C_4H_{11}N$	109-89-7	Diethylamine	0.58
$C_4H_{11}N$	78-81-9	Isobutylamine	0.73
$C_4H_{11}NO_2$	111-42-2	Diethanolamine	−1.43
C_5Cl_6	77-47-4	Hexachlorocyclopentadiene	5.04
$C_5H_4O_2$	98-01-1	Furfural	0.46

Continued

Formula	CAS Registry Number	Name	log K_{ow}
C_5H_5N	110-86-1	Pyridine	0.65
$C_5H_6O_2$	98-00-0	Furfuryl alcohol	0.28
C_5H_8	627-19-0	1-Pentyne	1.98
C_5H_9NO	872-50-4	N-Methylpyrrolidone	−0.54
$C_5H_9NO_2$	147-85-3	L-Proline	−2.54
C_5H_{10}	287-92-3	Cyclopentane	3.00
C_5H_{10}	109-67-1	1-Pentene	2.80
$C_5H_{10}N_2O_2S$	16752-77-5	Methomyl	0.60
$C_5H_{10}N_2O_3$	56-85-9	L-Glutamine	−3.15
$C_5H_{10}O$	96-22-0	Diethyl ketone	0.82
$C_5H_{10}O$	142-68-7	Tetrahydropyran	0.95
$C_5H_{10}O$	107-87-9	2-Pentanone	0.84
$C_5H_{10}O_2$	109-52-4	Pentanoic acid	1.39
$C_5H_{10}O_2$	109-60-4	Propyl acetate	1.24
$C_5H_{10}O_2$	108-21-4	Isopropyl acetate	1.02
$C_5H_{11}N$	110-89-4	Piperidine	0.84
$C_5H_{11}NO_2$	72-18-4	L-Valine	−2.26
$C_5H_{11}NO_2S$	63-68-3	L-Methionine	−2.00
C_5H_{12}	109-66-0	Pentane	3.39
C_5H_{12}	463-82-1	Neopentane	3.11
$C_5H_{12}NO_3PS_2$	60-51-5	Dimethoate	0.50
$C_5H_{12}O$	71-41-0	1-Pentanol	1.51
$C_6Cl_5NO_2$	82-68-8	Quintozene (PCNB)	4.64
C_6Cl_6	118-74-1	Hexachlorobenzene	5.44
C_6HCl_5	608-93-5	Pentachlorobenzene	5.17
C_6HCl_5O	87-86-5	Pentachlorophenol	5.18
$C_6H_2Cl_4$	634-66-2	1,2,3,4-Tetrachlorobenzene	4.54
$C_6H_2Cl_4$	95-94-3	1,2,4,5-Tetrachlorobenzene	4.63
$C_6H_2Cl_4$	634-90-2	1,2,3,5-Tetrachlorobenzene	4.63
$C_6H_2Cl_4O$	4901-51-3	2,3,4,5-Tetrachlorophenol	4.21
$C_6H_2Cl_4O$	58-90-2	2,3,4,6-Tetrachlorophenol	4.45
$C_6H_3Cl_3$	87-61-6	1,2,3-Trichlorobenzene	4.05
$C_6H_3Cl_3$	120-82-1	1,2,4-Trichlorobenzene	4.02
$C_6H_3Cl_3$	108-70-3	1,3,5-Trichlorobenzene	4.15
$C_6H_3Cl_3O$	15950-66-0	2,3,4-Trichlorophenol	3.61
$C_6H_3Cl_3O$	88-06-2	2,4,6-Trichlorophenol	3.69
$C_6H_4Cl_2$	541-73-1	1,3-Dichlorobenzene	3.52
$C_6H_4Cl_2$	95-50-1	1,2-Dichlorobenzene	3.38
$C_6H_4Cl_2$	106-46-7	1,4-Dichlorobenzene	3.45
$C_6H_4Cl_2O$	120-83-2	2,4-Dichlorophenol	3.17
$C_6H_4Cl_2O$	87-65-0	2,6-Dichlorophenol	2.64
C_6H_5Cl	108-90-7	Chlorobenzene	2.84
C_6H_5ClO	95-57-8	2-Chlorophenol	2.15
$C_6H_5NO_2$	98-95-3	Nitrobenzene	1.85
$C_6H_5NO_3$	88-75-5	2-Nitrophenol	1.77
$C_6H_5NO_3$	100-02-7	4-Nitrophenol	1.91
C_6H_6	71-43-2	Benzene	2.13

Formula	CAS Registry Number	Name	log K_{ow}
$C_6H_6Cl_6$	58-89-9	Lindane	3.55
C_6H_6O	108-95-2	Phenol	1.50
C_6H_7N	62-53-3	Aniline	0.90
$C_6H_8N_2$	111-69-3	Adiponitrile	−0.32
$C_6H_9N_3O_2$	71-00-1	L-Histidine	−2.52
C_6H_{10}	110-83-8	Cyclohexene	2.86
C_6H_{10}	693-02-7	1-Hexyne	2.52
$C_6H_{10}N_2O_2$	7491-74-9	Piracetam	−1.54
$C_6H_{10}O$	108-94-1	Cyclohexanone	0.81
C_6H_{12}	592-41-6	1-Hexene	3.40
C_6H_{12}	110-82-7	Cyclohexane	3.44
$C_6H_{12}O$	591-78-6	2-Hexanone	1.38
$C_6H_{12}O$	108-10-1	Methylisobutyl ketone	1.31
$C_6H_{12}O$	108-93-0	Cyclohexanol	1.23
$C_6H_{12}O_2$	142-62-1	Hexanoic acid	1.92
$C_6H_{12}O_2$	110-19-0	Isobutyl acetate	1.78
$C_6H_{12}O_2$	123-86-4	Butyl acetate	1.78
$C_6H_{13}N$	108-91-8	Cyclohexylamine	1.49
$C_6H_{13}NO_2$	73-32-5	L-Isoleucine	−1.76
$C_6H_{13}NO_2$	61-90-5	L-Leucine	−1.74
C_6H_{14}	110-54-3	Hexane	3.90
C_6H_{14}	75-83-2	2,2-Dimethylbutane	3.82
C_6H_{14}	79-29-8	2,3-Dimethylbutane	3.42
$C_6H_{14}N_2O$	621-64-7	N-Nitrosodipropyl amine	2.45
$C_6H_{14}O$	108-20-3	Diisopropyl ether	1.52
$C_6H_{14}O$	111-27-3	1-Hexanol	2.03
$C_6H_{14}O_2$	111-76-2	2-Butoxyethanol	0.80
$C_6H_{15}N$	121-44-8	Triethylamine	1.45
$C_6H_{15}NO_3$	102-71-6	Triethanolamine	−1.00
$C_7H_4Cl_4O_2$	2539-17-5	Tetrachloro-4-methoxyphenol	4.59
C_7H_5N	100-47-0	Benzonitrile	1.56
$C_7H_5N_3O_6$	118-96-7	Trinitrotoluene (TNT)	1.73
$C_7H_6ClN_3O_4S_2$	59-94-6	Chlorothiazide	−0.24
$C_7H_6N_2O_4$	121-14-2	2,4-Dinitrotoluene	1.98
$C_7H_6N_2O_4$	606-20-2	2,6-Dinitrotoluene	2.06
C_7H_6O	100-52-7	Benzaldehyde	1.48
C_7H_7ClO	59-50-7	4-Chloro-3-methylphenol	3.10
C_7H_8	108-88-3	Toluene	2.73
$C_7H_8N_4O_2$	58-55-9	Theophylline	−0.02
C_7H_8O	95-48-7	o-Cresol	1.98
C_7H_8O	108-39-4	m-Cresol	1.98
C_7H_8O	106-44-5	p-Cresol	1.97
$C_7H_{12}ClN_5$	122-34-9	Simazine	2.18
$C_7H_{13}N_3O_3S$	23135-22-0	Oxamyl	−0.47
C_7H_{14}	108-87-2	Methylcyclohexane	3.88
$C_7H_{14}N_2O_2S$	116-06-3	Aldicarb	1.13

Continued

Formula	CAS Registry Number	Name	$\log K_{ow}$
$C_7H_{14}O$	110-43-0	2-Heptanone	1.98
C_7H_{16}	142-82-5	Heptane	4.66
$C_7H_{16}O$	111-70-6	1-Heptanol	2.62
$C_7H_{17}O_2PS_3$	298-0202	Phorate	3.56
$C_8H_5Cl_3O_3$	93-76-5	2,4,5-T	3.31
$C_8H_6Cl_2O_3$	94-75-7	2,4-D	2.81
$C_8H_6Cl_2O_3$	1918-00-9	Dicamba	2.21
C_8H_7N	120-72-9	Indole	2.14
C_8H_7N	140-29-4	Benzeneacetonitrile	1.56
C_8H_8	100-42-5	Styrene	3.05
$C_8H_8Cl_3O_3PS$	299-84-3	Ronnel (Fenchlorphos)	4.88
C_8H_{10}	95-47-6	o-Xylene	3.12
C_8H_{10}	108-38-3	m-Xylene	3.20
C_8H_{10}	106-42-3	p-Xylene	3.15
C_8H_{10}	100-41-4	Ethylbenzene	3.15
$C_8H_{10}NO_5PS$	298-00-0	Parathion methyl	2.86
$C_8H_{10}N_4O_2$	58-08-2	Caffeine	−0.07
$C_8H_{10}O$	105-67-9	2,4-Dimethylphenol	2.35
$C_8H_{14}ClN_5$	1912-24-9	Atrazine	2.61
$C_8H_{14}N_4OS$	21087-64-9	Metribuzin	1.70
$C_8H_{16}NO_5P$	141-66-2	Dicrotophos	0.00
C_8H_{18}	111-65-9	Octane	5.15
$C_8H_{18}O$	111-87-5	1-Octanol	3.07
$C_8H_{19}N$	111-92-2	Dibutylamine	2.83
$C_8H_{19}O_2PS_2$	13194-48-4	Ethoprophos (Mocop)	3.59
$C_8H_{19}O_2PS_3$	298-04-4	Disulfoton	4.02
$C_9H_4Cl_3NO_2S$	133-07-3	Folpet	2.85
$C_9H_5Cl_3N_4$	101-05-3	Anilazine	3.00
$C_9H_6Cl_6O_3S$	959-98-8	α-Endosulfan	3.83
$C_9H_8Cl_3NO_2S$	133-06-2	Captan	2.35
$C_9H_8O_4$	50-78-2	Aspirin	1.19
$C_9H_9Cl_2NO$	709-98-8	Propanil	3.07
$C_9H_9N_3O_2$	10605-21-7	Carbendazim (MBC)	1.43
$C_9H_{10}BrClN_2O_2$	13360-45-7	Chlorobromuron	3.09
$C_9H_{10}Cl_2N_2O_2$	330-55-2	Linuron	3.20
$C_9H_{11}Cl_3NO_3PS$	2921-88-2	Chlorpyrifos	4.96
$C_9H_{11}NO_2$	63-91-2	L-Phenylalanine	−1.52
$C_9H_{11}NO_3$	60-18-4	L-Tyrosine	−2.66
C_9H_{12}	98-82-8	Isopropylbenzene (Cumene)	3.66
C_9H_{12}	108-67-8	Mesitylene	3.42
C_9H_{12}	95-63-6	1,2,4-Trimethylbenzene	3.70
$C_9H_{12}NO_5PS$	122-14-5	Fenitrothion	3.30
$C_9H_{12}N_2O_6$	58-96-8	Uridine	−1.98
$C_9H_{13}BrN_2O_2$	314-40-9	Bromacil	2.11
$C_9H_{13}ClN_2O_2$	5902-51-2	Terbacil	1.89
$C_9H_{13}ClN_6$	21725-96-2	Cyanazine	2.22
$C_9H_{13}N_3O_5$	65-46-3	Cytidine	−2.51

Formula	CAS Registry Number	Name	log K_{ow}
$C_9H_{16}ClN_5$	139-40-2	Propazine	2.93
$C_9H_{16}ClN_5$	5915-41-3	Terbuthylazine	3.06
$C_9H_{17}NOS$	2212-67-1	Molinate	3.21
$C_9H_{17}N_5S$	834-12-8	Ametryn	2.98
$C_9H_{18}N_2O_4$	57-53-4	Meprobamate	0.70
$C_9H_{19}NOS$	759-94-4	EPTC	3.21
$C_9H_{21}O_2PS_3$	13071-79-9	Counter (Terbufos)	4.48
$C_9H_{22}O_4P_2S_4$	563-12-2	Ethion	5.07
$C_{10}Cl_{12}$	2385-85-5	Mirex	5.28
$C_{10}H_7Cl$	90-13-1	1-Chloronaphthalene	4.24
$C_{10}H_7Cl$	91-58-7	2-Chloronaphthalene	4.14
$C_{10}H_7N_3S$	148-79-8	Thiabendazole	2.47
$C_{10}H_8$	91-20-3	Naphthalene	3.35
$C_{10}H_8O$	90-15-3	1-Naphthol	2.84
$C_{10}H_8O$	135-19-3	2-Naphthol	2.70
$C_{10}H_9Cl_4NO_2S$	2425-06-1	Captafol	2.51
$C_{10}H_9Cl_4O_4P$	22248-79-9	Tetrachlorvinphos	3.53
$C_{10}H_{10}O_4$	131-11-3	Dimethylphthalate	1.56
$C_{10}H_{11}ClO_3$	7085-19-0	Mecoprop	3.13
$C_{10}H_{11}F_3N_2O$	2164-17-2	Fluometuron	2.42
$C_{10}H_{12}$	119-64-2	Tetralin	3.49
$C_{10}H_{12}ClNO_2$	101-21-3	Chlorpropham	3.51
$C_{10}H_{12}N_2O_3S$	25057-89-0	Bentazone	2.80
$C_{10}H_{12}N_2O_5$	88-85-7	Dinoseb	3.69
$C_{10}H_{12}N_4O_2S_2$	94-19-9	Sulfaethidole	1.01
$C_{10}H_{12}N_4O_5$	58-63-9	Inosine	−2.10
$C_{10}H_{13}ClN_2$	6164-98-3	Chlordimeform	2.89
$C_{10}H_{13}Cl_2O_3PS$	97-17-6	Dichlofenthion	5.14
$C_{10}H_{13}NO_2$	94-12-2	Risocaine	2.40
$C_{10}H_{13}N_3$	1131-64-2	Debisoquin	0.75
$C_{10}H_{13}N_5O_4$	58-61-7	Adenosine	−1.23
$C_{10}H_{13}N_5O_4$	30516-87-1	Zidovudine (AZT)	0.05
$C_{10}H_{13}N_5O_5$	118-00-3	Guanosine	−1.89
$C_{10}H_{14}$	99-87-6	p-Cymene	4.10
$C_{10}H_{14}NO_5PS$	56-38-2	Parathion	3.83
$C_{10}H_{14}N_2$	54-11-5	Nicotine	1.17
$C_{10}H_{14}N_2O_5$	50-89-5	Thymidine	−1.17
$C_{10}H_{15}O_3PS_2$	55-38-9	Fenthion	4.09
$C_{10}H_{16}N_6S$	51481-61-9	Cimetidine	0.40
$C_{10}H_{19}N_5O$	1610-18-0	Prometon	2.99
$C_{10}H_{19}O_6PS_2$	121-75-5	Malathion	2.36
$C_{10}H_{21}NOS$	1114-71-2	Pebulate	3.84
$C_{10}H_{21}NOS$	1929-77-7	Vernolate	3.84
$C_{11}H_{10}$	90-12-0	1-Methylnaphthalene	3.87
$C_{11}H_{10}$	91-57-6	2-Methylnaphthalene	4.00

Continued

Formula	CAS Registry Number	Name	log K_{ow}
$C_{11}H_{11}ClO_3$	22131-79-9	Alclofenac	2.48
$C_{11}H_{12}Cl_2N_2O_5$	56-75-7	Chloramphenicol	1.14
$C_{11}H_{12}NO_4PS_2$	732-11-6	Phosmet	2.78
$C_{11}H_{12}N_2O_2$	73-22-3	L-Tryptophan	−1.06
$C_{11}H_{14}ClNO$	1918-16-7	Propachlor	2.18
$C_{11}H_{15}BrClO_3PS$	41198-08-7	Profenophos	4.68
$C_{11}H_{15}NO_2$	114-26-1	Propoxur	1.52
$C_{11}H_{15}NO_2S$	2032-65-7	Methiocarb	2.92
$C_{11}H_{16}N_2O_3$	7413-36-7	Nifenolol	1.28
$C_{11}H_{17}NO$	31828-71-4	Mexiletine	2.15
$C_{11}H_{17}O_4PS_2$	115-90-2	Fensulfothion	2.23
$C_{11}H_{21}NOS$	1134-23-2	Cycloate	4.11
$C_{11}H_{23}NOS$	2008-41-5	Butylate	4.15
$C_{12}Cl_{10}$	2051-24-3	DecachloroPCB	9.20
$C_{12}HCl_9$	52663-77-1	2,2',3,3'4,5,5',6,6'-PCB	8.16
$C_{12}H_2Cl_6O$	75198-38-8	1,2,3,6,9-PCDF	5.65
$C_{12}H_2Cl_8$	2136-99-4	2,2',3,3',5,5',6,6'-PCB	7.15
$C_{12}H_3Cl_5O_2$	40321-76-4	1,2,3,7,8-PCDD	6.64
$C_{12}H_3Cl_7$	52663-71-5	2,2',3,3',4,4',6-PCB	6.99
$C_{12}H_4Cl_6$	35065-27-1	2,2',4,4',5,5'-PCB	6.80
$C_{12}H_5Cl_3O_2$	39227-58-2	1,2,4-PCDD	7.47
$C_{12}H_5Cl_5$	18259-05-7	2,3,4,5,6-PCB	6.52
$C_{12}H_6Cl_2O$	5409-83-6	2,8-PCDF	5.65
$C_{12}H_6Cl_4$	35693-99-3	2,2',5,5'-PCB	6.09
$C_{12}H_7ClO_2$	39227-53-7	1-PCDD	5.05
$C_{12}H_7Cl_2NO_3$	1836-75-5	Nitrofen	4.64
$C_{12}H_7Cl_3$	15862-07-4	2,4,5-PCB	5.81
$C_{12}H_8Cl_2$	2050-68-2	4,4'-PCB	5.58
$C_{12}H_8Cl_6O$	60-57-1	Dieldrin	4.90
$C_{12}H_8O$	132-64-9	Dibenzofuran	4.12
$C_{12}H_8O_2$	262-12-4	Dibenzodioxin	4.37
$C_{12}H_8S$	132-65-0	Dibenzothiopene	4.38
$C_{12}H_9ClF_3N_3O$	273-14-13-2	Norflurazon	2.30
$C_{12}H_{10}$	92-52-4	Biphenyl	3.98
$C_{12}H_{10}Cl_2N_2$	91-94-1	3,3'-Dichlorobenzidine	3.51
$C_{12}H_{10}O$	90-43-7	2-Hydroxybiphenyl	3.09
$C_{12}H_{10}O$	92-69-3	4-Hydroxybiphenyl	3.20
$C_{12}H_{10}O$	101-84-8	Diphenyl ether	4.21
$C_{12}H_{11}ClN_2O_5S$	54-31-9	Furosemide	2.03
$C_{12}H_{11}Cl_2N_3O_2$	60207-31-0	Azaconazole	2.32
$C_{12}H_{11}N$	122-39-4	Diphenylamine	3.50
$C_{12}H_{11}NO_2$	63-25-2	Carbaryl	2.36
$C_{12}H_{11}N_3$	60-09-3	4-Aminoazobenzene	3.41
$C_{12}H_{12}N_2$	92-87-5	Benzidine	1.34
$C_{12}H_{12}N_2O_3$	50-06-6	Phenobarbital	1.47
$C_{12}H_{13}C_1N_2O$	3766-60-7	Buturon	3.00
$C_{12}H_{13}NO_2S$	5234-68-4	Carboxin	2.14

Formula	CAS Registry Number	Name	log K_{ow}
$C_{12}H_{14}Cl_3O_4P$	470-90-6	Chlorvinphos	3.81
$C_{12}H_{14}N_4O_2S$	3930-20-9	Sulfamethazine	0.28
$C_{12}H_{14}O_4$	84-66-2	Diethylphthalate	2.47
$C_{12}H_{15}NO_3$	1563-66-2	Carbofuran	2.32
$C_{12}H_{16}N_3O_3PS$	24017-47-8	Triazophos	3.55
$C_{12}H_{19}ClNO_3P$	299-86-5	Crufomate	3.42
$C_{12}H_{20}N_2O_3S$	3930-20-9	Sotalol	0.24
$C_{12}H_{21}N_2O_3PS$	333-41-5	Diazinon	3.81
$C_{12}H_{27}N$	102-82-9	Tributylamine	4.60
$C_{13}H_8F_2O_3$	22494-42-4	Diflunisal	4.44
$C_{13}H_{10}$	86-73-7	Fluorene	4.18
$C_{13}H_{16}F_3N_3O_4$	1582-09-8	Trifluralin	5.34
$C_{13}H_{19}N_3O_4$	40487-42-1	Pendimethalin	5.18
$C_{13}H_{20}N_2O_2$	59-46-1	Procaine	1.87
$C_{13}H_{21}NO_3$	5741-22-0	Morprolol	1.69
$C_{13}H_{21}N_3O$	51-06-9	Procainamide	0.88
$C_{13}H_{22}NO_3PS$	22224-92-6	Fenamiphos	3.23
$C_{14}H_8Cl_4$	72-55-9	p,p'-DDE	6.96
$C_{14}H_9ClF_2N_2O_2$	3567-38-5	Diflubenzuron	3.88
$C_{14}H_9Cl_2NO_5$	42576-02-3	Bifenox	4.48
$C_{14}H_9C_{15}$	50-29-3	p,p'-DDT	6.36
$C_{14}H_{10}$	85-01-8	Phenanthrene	4.52
$C_{14}H_{10}$	120-12-7	Anthracene	4.50
$C_{14}H_{10}Cl_2O_3$	34645-84-6	Fenclofenac	4.80
$C_{14}H_{10}F_3NO_2$	530-78-9	Flufenamic acid	5.25
$C_{14}H_{14}N_8O_4S_3$	25953-19-9	Cefazolin	−0.58
$C_{14}H_{18}N_4O_3$	738-70-5	Trimethoprim	0.91
$C_{14}H_{18}N_4O_3$	17804-35-2	Benomyl	2.12
$C_{14}H_{19}O_6P$	7700-17-6	Crotoxyphos	3.30
$C_{14}H_{20}$	15972-60-8	Alachlor	3.52
$C_{14}H_{20}N_2O_2$	34915-68-9	Bunitrolol	2.00
$C_{14}H_{20}N_2O_2$	21870-06-4	Pindolol	1.75
$C_{14}H_{21}N_3O_3$	4849-32-5	Karbutilate	1.66
$C_{14}H_{22}BrN_3O_2$	4093-35-0	Bromopride	2.83
$C_{14}H_{22}ClNO_2$	14556-46-8	Bupranolol	2.80
$C_{14}H_{22}N_2O$	137-58-6	Lidocaine	2.26
$C_{14}H_{22}N_2O_3$	29122-68-7	Atenolol	0.16
$C_{14}H_{22}N_2O_3$	6673-35-4	Practolol	0.79
$C_{14}H_{24}NO_4PS_3$	741-58-2	Bensulide	4.22
$C_{15}H_{12}N_2O_2$	57-41-0	Phenytoin	2.47
$C_{15}H_{14}N_4O_2S$	526-08-9	Sulfaphenazole	1.52
$C_{15}H_{15}ClN_2O_2$	1982-47-4	Chloroxuron	3.20
$C_{15}H_{18}N_4O_5$	50-07-7	Mitomycin-C	−0.40
$C_{15}H_{21}NO_2$	57-42-1	Meperidine	2.45
$C_{15}H_{22}ClNO_2$	51218-45-2	Metolachlor	3.13
$C_{15}H_{23}NO_2$	23846-70-0	Alprenolol	2.65

Continued

Formula	CAS Registry Number	Name	log K_{ow}
$C_{15}H_{23}NO_3$	6452-71-7	Oxprenolol	2.18
$C_{15}H_{23}N_3O_4S$	15676-16-1	Sulpiride	0.62
$C_{15}H_{24}N_2O_2$	94-24-6	Tetracaine	3.73
$C_{15}H_{24}N_2O_4S$	51012-32-9	Tiapride	0.90
$C_{15}H_{25}NO_3$	37350-58-6	Metoprolol	1.88
$C_{16}H_{10}$	206-44-0	Fluoranthene	5.20
$C_{16}H_{10}$	129-00-0	Pyrene	5.00
$C_{16}H_{10}N_2O_2$	482-89-3	Indigo	3.72
$C_{16}H_{13}ClN_2O$	439-14-5	Diazepam	2.02
$C_{16}H_{14}O_3$	36330-85-5	Fenbufen	3.12
$C_{16}H_{14}O_3$	22071-15-4	Ketoprofen	3.12
$C_{16}H_{15}C_{13}O_2$	72-43-5	Methoxychlor	4.95
$C_{16}H_{16}N_2O_4$	13684-56-5	Desmedipham	3.39
$C_{16}H_{16}N_2O_4$	13684-63-4	Phenmedipham	3.59
$C_{16}H_{19}N_3O_4S$	69-53-4	Ampicillin	−1.13
$C_{10}H_{20}N_4O_6$	57998-68-2	Diaziquone	−0.02
$C_{16}H_{21}NO_2$	525-66-6	Propanolol	3.09
$C_{16}H_{22}O_4$	84-74-2	Dibutyl phthalate	4.72
$C_{16}H_{25}NO_3$	54-32-0	Moxisylyte	3.17
$C_{17}H_{12}Cl_2N_2O$	60168-88-9	Fenarimol	3.60
$C_{17}H_{16}ClFN_2O_2$	62666-20-0	Progabide	2.97
$C_{17}H_{19}ClN_2S$	50-53-3	Chlorpromazine	5.35
$C_{17}H_{19}NO_3$	57-27-2	Morphine	0.76
$C_{17}H_{20}N_2S$	58-40-2	Promazine	4.55
$C_{17}H_{21}NO$	58-73-1	Diphenhydramine	3.40
$C_{17}H_{21}NO_2$	15299-99-7	Napropamide	3.36
$C_{17}H_{21}NO_4$	50-36-2	Cocaine	2.30
$C_{17}H_{23}NO_3$	51-55-8	Atropine	1.83
$C_{18}H_{10}$	203-12-3	Benzo[ghi]fluoranthene	6.63
$C_{18}H_{12}$	218-01-9	Chrysene	5.86
$C_{18}H_{12}$	92-24-0	Naphthacene	5.76
$C_{18}H_{12}$	56-55-3	Benz[a]anthracene	5.91
$C_{18}H_{19}F_3N_2S$	146-54-3	Fluopromazine	5.19
$C_{18}H_{21}NO_3$	76-57-3	Codeine	1.14
$C_{18}H_{22}O_2$	53-16-7	Estrone	3.13
$C_{18}H_{25}N_3O_5$	66203-00-7	Carocainide	1.38
$C_{18}H_{28}N_2O_4$	37517-30-9	Acebutolol	1.77
$C_{18}H_{29}NO_2$	38363-40-5	Penbutolol	4.15
$C_{19}H_{16}ClNO_4$	53-86-1	Indomethacin	4.27
$C_{19}H_{20}N_2O$	73931-96-1	Denzimol	3.40
$C_{19}H_{20}N_2O_2$	50-33-9	Phenylbutazone	3.16
$C_{19}H_{20}O_4$	85-68-7	Benzylbutyl phthalate	4.91
$C_{19}H_{23}N_3$	33089-61-1	Amitraz	5.50
$C_{19}H_{24}N_2$	50-49-7	Imipramine	4.80
$C_{19}H_{24}N_2OS$	60-99-1	Levomepromazine	4.68
$C_{19}H_{25}N_5O_4$	63590-64-7	Terazosin	−0.38
$C_{19}H_{28}O_2$	58-22-0	Testosterone	3.32

Formula	CAS Registry Number	Name	log K_{ow}
$C_{20}H_{12}$	50-32-8	Benzo[a]pyrene	6.35
$C_{20}H_{12}$	198-55-0	Perylene	6.25
$C_{20}H_{13}ClF_3NO_3$	53966-34-0	Floxacrine	3.46
$C_{20}H_{23}N$	50-48-6	Amitriptyline	5.04
$C_{20}H_{24}N_2O_2$	50-54-2	Quinidine	2.64
$C_{20}H_{25}N_3O$	50-37-3	LSD-25	2.95
$C_{21}H_{20}C_{12}O_3$	52645-53-1	Permethrin	6.50
$C_{21}H_{23}ClFNO_2$	52-86-8	Haloperidol	3.36
$C_{21}H_{24}F_3N_3S$	117-89-5	Trufluoperazine	5.03
$C_{21}H_{26}ClN_3OS$	58-39-9	Perphenazine	4.20
$C_{21}H_{26}N_2S_2$	50-52-2	Thioridazine	5.90
$C_{21}H_{27}NO$	76-99-3	Methadone	3.93
$C_{21}H_{27}N_3O_3$	252-06-7	Nicainoprol	1.63
$C_{21}H_{29}NO$	514-65-8	Biperiden	4.25
$C_{21}H_{30}O_5$	50-23-7	Hydrocortisone	1.61
$C_{22}H_{12}$	191-24-2	Benzo[ghi]perylene	6.90
$C_{22}H_{14}$	53-70-3	Dibenz[a,h]anthracene	6.75
$C_{22}H_{19}Br_2NO_3$	52918-63-5	Deltamethrin	6.20
$C_{22}H_{25}N_3O_4S$	31883-05-3	Moricizine	2.98
$C_{22}H_{26}F_3N_3OS$	69-23-8	Fluphenazine	4.36
$C_{22}H_{26}N_2O_4S$	42399-41-7	Diltiazem	2.70
$C_{22}H_{29}N_3S_2$	1420-55-9	Thiethylperazine	5.41
$C_{22}H_{29}N_7O_5$	53-79-2	Puromycin	0.03
$C_{22}H_{30}N_2O_2S$	56030-54-7	Sufentanil	3.95
$C_{24}H_{12}$	191-07-1	Coronene	6.50
$C_{24}H_{38}O_4$	117-81-7	Bis(2-ethylhexyl)phthalate	7.88
$C_{25}H_{22}ClNO_3$	51630-58-1	Fenvalerate	6.20
$C_{25}H_{38}O_5$	79902-63-9	Simvastin	4.68
$C_{27}H_{38}N_2O_4$	52-53-9	Verapamil	3.83
$C_{28}H_{29}F_2N_3O$	2062-78-4	Pimozide	6.30
$C_{32}H_{32}O_{13}S$	29767-20-2	Teniposide	1.24
$C_{34}H_{46}ClN_3O_{10}$	35846-53-8	Maytansine	1.99
$C_{37}H_{67}NO_{13}$	114-07-8	Erythromycin	2.54
$C_{41}H_{64}O_{13}$	71-63-6	Digitoxin	2.83
$C_{49}H_{76}O_{20}$	17575-22-3	Lanatoside-C	0.07

LOGKOW on diskette may be obtained from: Sangster Research Laboratories
Suite 402
3475 de la Montagne
Montreal, Quebec, Canada
H3G 2A4

LOGKOW on diskette and as an on-line version are available from: Technical Database Services, Inc.
135 West 50th Street
New York, NY 10020-1201
USA

INDEX